Neues verkehrswissenschaftliches Journal

Ausgabe 16

Capacity Research in Urban Rail-Bound Transportation with Special Consideration of Mixed Traffic

(Leistungsuntersuchungen im städtischen schienengebundenen Verkehr unter besonderer Berücksichtigung des Mischverkehrseinflusses)

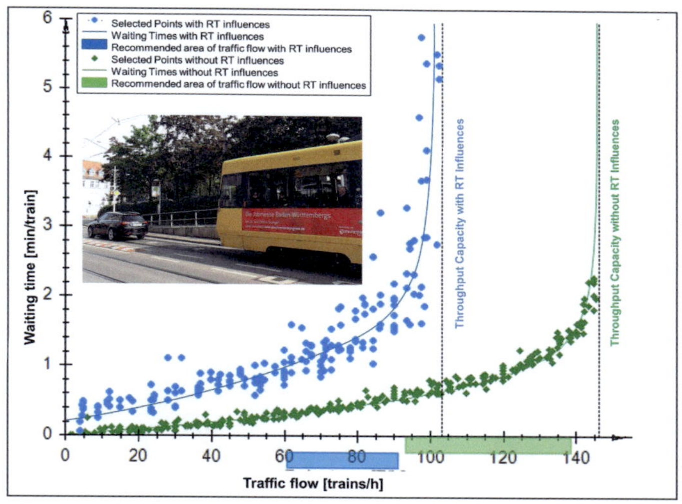

DFG – Forschungsprojekt (MA 2326/13-1)

Prof. Dr.-Ing. Ullrich Martin

M.Sc. Di Liu

Institut für Eisenbahn- und Verkehrswesen der Universität Stuttgart

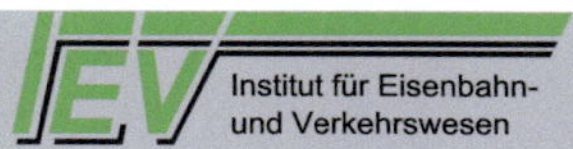

Titelbild: Ullrich Martin, Di Liu

Herstellung und Verlag: Books on Demand GmbH, Norderstedt

Printed in Germany

ISBN 978-3-7431-8067-3

Preface

In 2013, the research project with the topic "Capacity Research in Urban Rail-Bound Transportation with Special Consideration of Mixed Traffic" was proposed and granted by the Deutsche Forschungsgemeinschaft (DFG). The research project was completed in 2015, in which the application of capacity research for the evaluation of rail-bound systems is applied to urban rail-bound transport with external influences caused by road traffic in urban mixed traffic zones.

In this completed research project, a simulation model for urban rail-bound transport with consideration of road traffic influences can be built through two developed approaches. The modeling approach—through modeling the road traffic as urban rail-bound transport in a simulation model, and the distribution approach—using mathematical distributions to convert the road traffic influences to perturbations in the simulation model, the influences of road traffic on urban rail-bound transport can be determined for capacity research. One important intermediate result of capacity research, the model function of the waiting time function, is further improved to be adapted to the urban rail-bound system with consideration of external influence (caused by road traffic in urban mixed traffic zones). In addition, a developed algorithm with an event-driven system for urban mixed traffic can be used for the determination of the waiting time function and the throughput capacity for the evaluation of urban rail-bound transport with consideration of mixed traffic, which is also applied for the preliminary assessment of the significances of road traffic influences at each mixed traffic zone in a whole investigated area.

With the results of this project, the method and application of capacity research for the macroscopic and mesoscopic evaluation is supplied and expanded to include urban mixed traffic, so that the quality of operation and the operational performance of urban rail-bound transport with road traffic influences at mixed traffic zone can be further evaluated.

Stuttgart, December 2016

Ullrich Martin

Table of Content

List of Figures

List of Tables

1 Introduction

Nowadays, the possibility for the construction of new railway infrastructure and the extension of existing infrastructure has declined after decades of development. However, economic development and an accompanying increase in the number of vehicles on the roads, especially in urban areas, has been observed. Important questions for the future research of urban mixed traffic include how to use the existing infrastructure as intensively and as reasonably as possible, while maximizing the capacity of the infrastructure without reducing the required operational quality. A key ingredient in doing this is by analyzing the waiting time function that reflects the relationship between the quality of operation (waiting time) and the capacity. The waiting time function can be derived from the results of simulations of stochastic influenced timetables with stepwise-varied traffic flows of trains (different compaction percentages of the given operating program) due to the operational hindrances between trains.

In urban mixed traffic zones, urban rail-bound transport (URT) interacts with road traffic (ROT), which includes motorized road traffic as well as non-motorized road traffic. Due to the inherent complexity of urban mixed traffic zones, existing studies in the field of capacity research have rarely considered the external influences caused by road traffic. These influences that impact the simulations lead to a deviation of the existing waiting time function. Hence, the results of capacity research without consideration of road traffic in mixed traffic zones are inaccurate. In order to improve the accuracy of capacity research, algorithms for capacity research of urban rail-bound transport were developed in this research project with special consideration given to the influences of road traffic in mixed traffic zones.

In this research project supported by the Deutsche Forschungsgemeinschaft with the reference number of "MA 2326/13-1", methods for the capacity research of urban rail-bound transport were investigated with special consideration of the road traffic influences in mixed traffic zones. A general introduction to capacity research and two main methods of capacity research for rail-bound transport are presented in Chapter 2. To model the urban rail-bund transport with road traffic influences to a sufficient accuracy level without redundancy, the significant influential parameters were identified and studied with a sensitivity analysis in Chapter 3, which were based on data collected from on-site measurements. Two approaches (the modeling approach and the distribution approach) were developed to execute the capacity research with consideration of road traffic influences. With regards to the modeling approach, the road traffic was modeled comparable to trains in simulation tools to reflect the influences of road traffic, as described in Chapter 4. With regards to the distribution approach, the influences of road traffic on the urban rail-bound transport were modeled as operational perturbations in the simulation process with the help of a simulation tool, as described in Chapter 5. The capacity research was carried out with the simulation method assisted with the two approaches. Because of the additional external influences of road traffic, the existing waiting time function cannot fit the simulation results properly. An adapted waiting time function for capacity research with consideration of road traffic influences was further improved and is described in Chapter 6. Furthermore, an algorithm for determining the adapted waiting time function for capacity research was developed. Finally, the results of capacity research of the two approaches were compared and evaluated in Chapter 7.

2 Capacity Research of Rail-Bound Transport

In this chapter, a general summary is made about capacity research for the rail-bound transport of the railway system. Firstly, Subchapter 2.1 manifests the basic concept of capacity research for the railway system. The relevant basic terms will be introduced in Subchapter 2.2. Two major methodologies for capacity research of rail-bound transport, which are the analytical method and the simulation method, will be presented in Subchapter 2.3.

2.1 Overview

Capacity research is one of the most important methods for the validation of timetables on an existing or planned infrastructure, for adequate railway capacity design of the infrastructure, as well as the performance evaluation for railway operation [Pachl 2014]. In railway operation, capacity research is a significant issue that evaluates the operating performance of the railway systems. The operating performance is generally considered to be the quality of operation and the number of trains per time unit (hour) running in an investigated railway infrastructure network. The quality of operation usually refers to the waiting time (operational delays) in operation. Therefore, the capacity is inversely proportional to the quality of operation. One goal of capacity research is to improve the performance through optimizing the existing infrastructure and the operating program. Determination of the quality of operation for an investigated area with a concrete timetable and determination of recommended area of traffic flow with a rough operating program can both be achieved with railway capacity research.

2.2 Relevant Terms

Relevant terms of capacity research for railway systems in this research project are defined as following:

- ***Traffic flow [trains/h]*** is the number of trains or train paths during a given investigation period within the investigated area [DB Netz AG 2008].

- ***Operating program*** is the comprehensive description of the performance and requirements of the railway operation, including the number of train runs, the train properties, their structure and sequence, as well as the temporal allocation of the train runs [DB Netz AG 2008].

- **_Train mixture_** is the structure of the operating program, which is the rough operating program including the characteristics of the various train groups[1] in the model and the ratio of the number of trains in each train group.

- **_Maximum capacity_** is a theoretical value and doesn't meet the requirements of operational quality. It allows for an unlimited congestion situation without keeping the structure of the operating program. It corresponds to the theoretical maximum number of trains and shunting movements on the investigated infrastructure within a given time period [DB Netz AG 2008].

- **_(Maximum) throughput capacity (DS LF)_** is the average traffic flow of all possible maximum traffic flows per time unit (hour) in the static phase of the operating procedure with various train sequences of a defined rough operating program (train mixture). The throughput capacity (incoming traffic flow = outgoing traffic flow) has to keep the structure of the defined operating program (train mixture) on a given infrastructure. A further increase of traffic flow leads to a slowing growth trend or a change of the operating program (train mix) of the outgoing traffic flow [Chu 2014].

- **_Investigated area_** is a defined part of the railway infrastructure on which the capacity research will be carried out.

- **_Utility factor of capacity_** is the amount of consumed capacity, which is calculated as the actual traffic flow divided by the throughput capacity.

- **_Recommended area of traffic flow_** is the range of traffic flow to reach a customer-friendly quality of operation with the optimum utilization of a given infrastructure. It is located between the lower limit with the minimum value of relative sensitivity function of the waiting time, and the higher limit with the maximum value of the so-called traffic energy function ([Hertel 1992], [Schmidt 2009] & [Chu 2014]).

---

[1] Train group: Group of trains with same or similar characteristics, which operate on the same or similar routes.

2.3 Methods of Capacity Research

There are various methods of capacity research, for the description of the operational quality and the further operating performance, the waiting time is one of the most important parameters of capacity research. Moreover, an intermediate result of capacity research is the throughput capacity for a specific operating program on a given railway infrastructure and during a specific time period. There are basically two methods to determine the waiting time and the throughput capacity for capacity research:

- Analytical method
- Simulation method

2.3.1 Analytical Method

The analytical method is a common method for conducting capacity research and is based on mathematic analysis. Using this method allows for the calculation of the capacity of the railway infrastructure including the railway lines and railway nodes (railway stations), and waiting time by means of mathematical expressions given the infrastructure and the characteristics of the operating program ([Potthoff 1969], [Schwanhäußer 1978], [Kontaxi & Ricci 2010] and [Pachl 2014]). The infrastructure and timetable (operating program) are modeled with suitable mathematic models, such as parallel servers and the probability distribution. For the analytical method, the basic theories are the queuing theory and the probability theory [Potthoff 1972]. The investigated area of the railway system can be modeled as a corresponding queuing system. Accordingly, the waiting time can be derived by using mathematical models.

For the analytical method, the railway operation process can be described mathematically either by the deterministic expression or the stochastic expression. The analytical method is based on certain assumptions and an investigated operating program by using mathematical methods to assess the capacity of the railway lines and nodes [Liu 2011]. From the mathematical point of view, the stochastic expression is utilized with unknown quantities that are mutually dependent on each other [Kontaxi & Ricci 2010]. On the other hand, the deterministic expression is used with unknown quantities that are mutually independent of each other [Kontaxi & Ricci 2010].

With the analytical method, the infrastructure is studied and calculated with the track lines and railway nodes, which are modeled with suitable queuing systems as single servers or multi-servers. For track lines subdivided into different sections, single severs are used for each direction when applying the analytical method. In addition, in order to measure the capacity of a railway node, it is necessary to analyze the structures of the node in the railway networks as well as the smaller infrastructure elements that make up a railway node.

The railway nodes are the points of the network connections where multiple railway lines are linked. The structure of a node consists of two components [Pachl 2016]: the set of tracks (Gleisgruppen) and the set of conflicting sub routes (Teilfahrstraßenknoten). The set of tracks in a node is modeled as multi-servers of a queuing system. Comparably, because the conflicting sub routes in the node are exclusive of each other, its characteristics can be modeled as single servers in a queuing system. For complex railway nodes, many sets of conflicting sub routes are assembled to be modeled as a multi-resource queue (see [Omahen 1977], [Green 1984] and [Nießen 2008]). Through mathematical analysis, an expected value of the waiting time and the line exploitation rate can be ascertained deterministically.

Furthermore, the operating program is described mathematically with the analytical method. When using the queuing system, the trains are treated as the customers.. The information from the operating program can be described as a random variable, which is the time interval between the arrival times of two trains, using a suitable distribution (or its variance in the simplest case). The occupation times of a train refer to the service time in the server of a queuing system, which is regarded as a random variable with a proper probability distribution (or its variance in the simplest case). Therefore, the operating program can be represented with a suitable mathematical model.

[International Union of Railways (UIC) 2004] describes the compression method to determine the maximum capacity with the time-distance diagram. Through compressing the blocking time stairways and keeping the minimum line headway and the structure of the trains, the consumed capacity of an infrastructure can be derived. However, this compression method as described in [International Union of Railways (UIC) 2004] can only be used under very large restrictions [Lindner 2011]. The line capacity utilized by an operating program can be simply visualized by compressing the blocking time stairways as close as possible together without any buffer time and with keeping the sequence of trains unchanged [Pachl 2016]. According to [Wendler 2002], it is comprehensible that the evaluation of the railway lines capacity research can be carried out using the compression method.

With its mathematical formulas and algebraic expressions, the analytical method is mainly utilized to identify the preliminary resolutions and reference values for capacity research [Rossetti 2009]. It is helpful to use the analytical method in relatively simple situations, which may require more effort and time to simulate than to solve the problem [Pachl 2014]. It is useful to apply the analytical method for relatively simple, homogeneous track arrangements and separated railway infrastructure sections to evaluate the performance of such restricted areas [Rossetti 2009]. The analytical method is designed to provide efficient results based on scant aggregated input data for evaluating the railway capacity on a homogeneous investigated railway infrastructure with a rough operating program instead of detailed timetables.

2.3.2   Simulation Method

Simulation is the modeling of a real object or process based on their considered characteristics. Simulation can be utilized to study and evaluate a real object or process instead of using the original real object or process [Siefer 2014].The simulation method is realized in a computer-based model of a railway system instead of the real operation [Pachl 2014].

A computer-based model can be set up through determining various attributes of the real system with the help of simulation tools like RailSys[2], which is introduced in [Siefer 2014]. In a virtual laboratory for a railway system, the infrastructure and timetables (based on the operating program) can be set and adjusted in various ways according to what is required. The model can be used to study and evaluate the real railway system without much effort of time and cost. It has few limitations and is relatively more cost-efficient than a physical experimental model in a real railway system for research and planning [Siefer 2014].

---

[2] RailSys: Rail Management Consultants GmbH (RMCon), Hannover. RMCon [2010]

The simulation method can be used for evaluating the operational performance of a railway system, optimizing and validating the operational performance, and planning timetables in a railway system. The railway system consists of various characteristics of the infrastructure including the tracks, switches, signaling systems, station (node) and lines; as well as the rolling stock [Siefer 2014]. In addition, it also includes the operation of trains with the rules on the infrastructure based on the timetable (operating program), and even the operational perturbations as well as dispatching with an appropriate model in the simulation tool. The model using the simulation method is built much closer to reality and higher accuracy of results is obtained. Capacity research can be carried out through simulating a series of timetables with stepwise-varied traffic flows, the results of the evaluation for an operating program or a specific timetable on a given infrastructure and the quality of operation can be obtained with the simulation method and can be further analyzed for optimization.

According to processing techniques with the help of simulation tools, the simulation method can be classified into two types:

- Asynchronous simulation
- Synchronous simulation

The difference between these two types is in the modeling of the railway operations of the trains. In asynchronous simulation, the trains are based on their priority and the temporal sequence in succession without changing the blocking time stairways (the position of the train paths can be changed but not their slopes in the time-distance-diagram). All of the trains in asynchronous simulation are modeled in the whole process with the help of simulation tool such as LUKS[3].

In comparison, with synchronous simulation all the trains are simulated simultaneously (synchronously) at each temporal point. The processes of railway operation in synchronous simulation are simulated in real temporal sequences with the help of simulation tools such as RailSys or OpenTrack[4]. The results of synchronous simulation

---

[3] LUKS: The software of railway simulation tool for research of an integrated system of railway operations, VIA Consulting & Development GmbH, Aachen. Janecek & Weymann [2010a] and Janecek et al. [2010b]

[4] OpenTrack: OpenTrack Railway Technology GmbH, Zürich. Michl & Sojka [2014]

are relatively closer to the reality of the operation process, whereas asynchronous simulation is primarily used for scheduling purposes. Nowadays, current simulation tools increasingly combine synchronous and asynchronous approaches.

In this research project, the key point is to evaluate the capacity of the urban rail-bound transport under the conditions of urban mixed traffic. Therefore, consideration will be given to the influences caused by the individual vehicles and pedestrians in urban mixed traffic zones with the aid of the simulation method; these influences are complex and stochastic. For a railway system, [Potthoff 1972] has already proposed the basic stochastic influences in the operational process of railway, which lead to a deviation between the planned and the actual operations. Stochastic influences are regarded as the disturbances randomly caused during railway operation.

The waiting time function is an important intermediate result of simulation method. It was the basis for developing the theory for capacity research of double-track railway lines, while determining the recommended area of traffic flow. It was first proposed by Hertel and Ludwig [Hertel et al. 1987]. The determination of the recommended area of traffic flow with the waiting time function is described in [Hertel 1992]. The theory of [Hertel 1992] allows the recommended area of traffic flow to be derived directly from the waiting time function. To obtain a reasonable and plausible result, the waiting time function has to be determined with sufficient accuracy. In the simulation method, the simulation model of the investigated area has to be suitably modeled. Moreover, the operating program or a scheduled timetable with train types and train mixture is also needed for the basis for the simulation.

Improvements to approximating the waiting time function with a linear approach was put forward and firstly used in the simulation method of railway operation [Bosse et al. 1995]. In order to carry out capacity research with the simulation method, the software PULEIV[5] was developed to work with other simulation tools to create stochastically-influenced timetables with stepwise-varied traffic flows of trains with consideration of the arrival distribution based on the operating program [Martin et al. 2011]. When the stepwise-created stochastic timetables are simulated, the recommended

---

[5] Programm zur Untersuchung des Leistungsverhaltens von Eisenbahninfrastrukturen, IEV, Stuttgart. Martin et al. [2011]

area of traffic flow can be determined by means of the fitted curve with the waiting time function. PULEIV was successfully applied in the research project by [Martin & Breuer 2008].

[Schmidt 2009] described the further development of capacity research with the simulation method based on previous research. In her dissertation, she describes a method for determining the maximum capacity and the waiting time function using a logarithmic approximation approach.

Further development on the creation of stochastic influenced timetables with different traffic flows based on the operating program was described in [Martin & Chu 2013] and [Martin & Chu 2012]. In this research, a new approach for creating more realistic stochastic-influenced timetables through time slice was pursued, which is needed for the simulation method. Furthermore in Chu's dissertation, a new approach was developed to determine the throughput capacity while keeping the structure of operating program unchanged and considering a transient phase in the simulation process; this successfully improved the waiting time function with three parameters (see Figure 2-1). The adaptability of the waiting time function and measures to enhance the stability of the approximation method were further investigated in [Chu 2014].

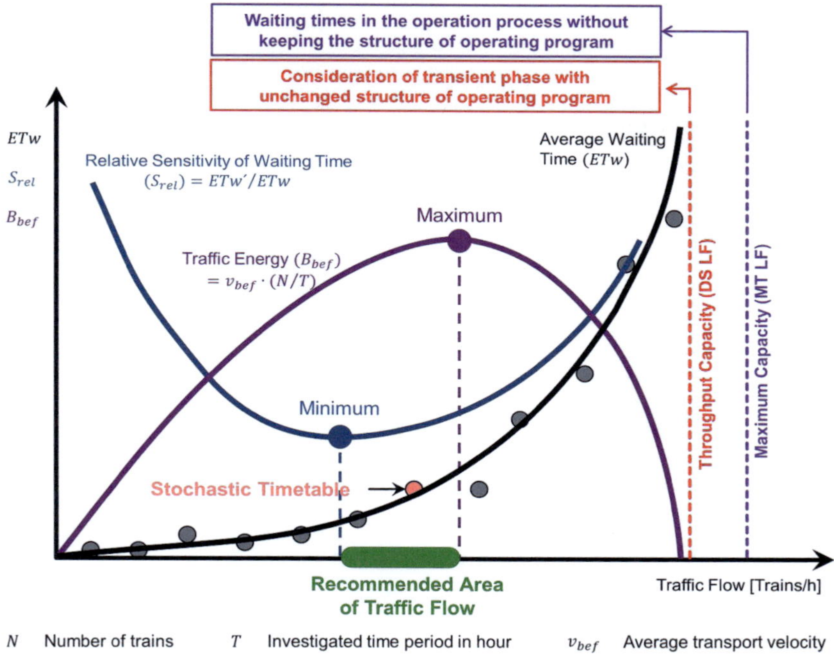

Figure 2-1: Evaluation of Performance for Capacity Research (Source: [Chu 2014] and [Li 2015])

With the simulation method, the intermediate results such as the transport time, entry and exit traffic flows, as well as the waiting time can be derived. Based on the method in [Hertel 1992], the recommended area of traffic flow for capacity research can be described. This method is based on the waiting time function with various parameters to express the relationship between capacity and quality of operation. These various parameters are determined by the given operating program and infrastructure.

Furthermore, based on the theory by [Schmidt 2009], the throughput capacity can be determined with consideration of a transient phase through the relationship between the entry traffic flow and the exit traffic flow [Chu 2014]. The throughput capacity can be regarded as the maximum capacity (the trains maximally running in the investigated area) with an unchanged structure of the operating program (train mixture). There is an important variable η needed for determining the waiting time function. It is indicated as the utility factor of capacity on a given infrastructure, which is the amount of consumed capacity. The utility factor of capacity results from the relationship of the expected value (average) of the service time (occupation time) of trains in the server of queuing system (ET_b) and the expected value (average) of time interval between arrival times of two trains (ET_a). The actual number of arrival trains per time unit (hour) can be represented by $1/ET_a$, which is the actual traffic flow of the given infrastructure. $1/ET_b$ is the average service rate (in the queuing system), which can represent the throughput capacity. Therefore, the utility factor of capacity η can be calculated as:

$$\eta = \frac{\frac{1}{ET_a}}{\frac{1}{ET_b}} = \frac{ET_b}{ET_a} = \frac{Traffic\ flow}{Throughput\ capacity} \qquad (\text{2-1})$$

With:

ET_a Expectancy value of time interval between the arrival times of two trains.

ET_b Expectancy value of the service time in the server of the queuing system.

The black curve in Figure 2-1 is the fitted curve of the discrete waiting times. Based on the Monte-Carlo-Method for simulation, the timetables with stepwise-varied traffic flows are simulated. Accordingly, the corresponding waiting times are assembled and fit to a curve based on the waiting time function, which will be introduced in Subchapter 6.2 of this report.

[Ludwig 1990] proposed the first model function of the waiting time function for research of railway track lines. In [Schmidt 2009], the method for determining the waiting time function with the simulation method was developed. Further development by [Chu 2014] proposed a new model function of the waiting time function for capacity research using three parameters a_1, b_1 and c_1 (see Subchapter 6.2.3). These parameters can be derived from the results of the simulation and from fitting the curve.

Through the derived waiting time function with the fitted parameters, the recommended area of traffic flow can be determined from two limit points from two further derived curves(see [Hertel 1992], [Martin 2014] and [Pachl 2016]):

- The lower limit of the recommended area of traffic flow is given by minimum value of the relative sensitivity of the waiting time function (2-2), which is the quotient of the first derivative of the waiting time function and the waiting time function itself.

$$S_{rel}(\eta) = \frac{ETw'(\eta)}{ETw(\eta)} \qquad (\text{2-2})$$

With:

$S_{rel}(\eta)$ Relative sensitivity of the waiting time function.

$ETw(\eta)$ (statistical) Expectancy value waiting time function.

$ETw'(\eta)$:   The first derivation of the waiting time function.

η Utility factor of capacity (traffic flow / throughput capacity).

- The upper limit is the maximum value of the traffic energy function (2-3), which is the product of the number of trains and the average speed per train.

$$B_{bef}(\eta) = \frac{N \cdot v_{bef}}{T} = \frac{\eta}{1 + \frac{ETw(\eta)}{ET_B}} \qquad (\text{2-3})$$

With:

$B_{bef}(\eta)$ Traffic energy function.

N Number of trains.

v_{bef} Average speed of the trains.

T Time interval.

$ETw(\eta)$ (statistical) Expectancy value waiting time function.

ET_B Average transport time.

η Utility factor of capacity (traffic flow / throughput capacity).

The stochastic timetables with different densities of traffic flows in the investigated area that are located within the recommended area of traffic flow represent the optimal relationship of capacity and quality of operation. The timetables with traffic flows lower than the lower limit of recommended area of traffic flow lead to a moderate increase of waiting time, which means too much free space has been left on the railway infrastructure. The timetables with traffic flows located to the right of the upper limit of the recommended area of traffic flow are timetables where the waiting time increases quickly with a slight increase of the number of trains. This case means that a loss of customers may occur due to the long waiting time (delays). The quality of the operation goes down, which is followed by a disadvantage of competition compared to other means of transport.

3 External Influences on Urban Rail-Bound Transport in Urban Mixed Traffic Zone

3.1 Urban Mixed Traffic

3.1.1 Overview

An urban mixed traffic zone is an area in which not only urban rail-bound transport operates but also other mutually interfering types of transport. These types of transport are collectively called urban mixed traffic. In this research project, the goal is to evaluate the operating performance and operational quality of the urban rail-bound transport under the conditions of urban mixed traffic. The basic categories to be studied in this research project are urban rail-bound transport and road traffic. Road traffic can be categorized into two different modes:

- Motorized road traffic
- Non-motorized road traffic

Motorized road traffic includes individual vehicles such as heavy vehicles (trucks), private cars, motorcycles, and public vehicles such as buses. Non-motorized road traffic includes pedestrians and bicycles.

The topology structures and the interactions between urban rail-bound transport and road traffic in an investigated mixed traffic zone can be complex and diverse. Basically, there are two types of mixed traffic zones:

- Level crossings
- Shared roads

A level crossing is a major type of urban mixed traffic zone in this research project. It is an intersection where an urban rail-bound line crosses a road or path at the same level. A level crossing is one of the most common types of mixed traffic zones. It allows for both urban rail-bound transport and road traffic to pass through it, and is typically controlled by traffic control signals (traffic lights).

A shared road is the other major type of mixed traffic zone discussed in this research project. On a shared road, the rail-bound tracks for urban rail-bound transport share a road with the urban roadways and even the bikeways (bicycle lanes. All of urban mixed traffic including urban rail-bound transport and road traffic run in-line in the same direction.

Theoretically, road traffic cannot influence urban rail-bound transport if urban rail-bound transport has the absolute priority. However in reality, there is regulated maximum time duration of the red light phase for road traffic at level crossings. A fundamental cause of road traffic influences at level crossings is the regulated limitation of the red light phase. In addition, the minimum green light phase of the traffic control signal for road traffic is also restricted. From another point of view, in mixed traffic zones, the different means of road traffic have different behaviors, such as speed, deceleration, and acceleration.

In general, roadways have speed limits and the different types of road traffic should drive in accordance with the corresponding speeds. Moreover, when the road traffic is comprised of different modes, such as bicycles or trucks, the speed limits for different road traffic modes are different. Furthermore, since there are traffic control signaling systems, the road traffic influences on urban rail-bound transport are slight at level crossings.

However for shared roads, the different speed limitations of the different road traffic modes effect the running times of the road traffic traveling along the shared road. The attributes of the shared road may further influence the running time of urban rail-bound transport if the road traffic traveling in front of it traveling slower, especially if there is road traffic congestion

Normally, the dynamics of vehicle movements between road traffic and urban rail-bound transport are different. However, due to the interactions between them along the shared road, the running time and the waiting time of urban rail-bound transport has to be in accordance with the congested road traffic traveling in front of the urban rail-bound transport vehicle: the speed of the road traffic in front of the urban rail-bound transport is significant. Therefore, the influences of the different dynamics of urban rail-bound transport and road traffic along the shared road exist but are not significant; these influences are ignored in this research project.

Relatively, pedestrian movements are much more complex than private cars, but in general, pedestrians also follow the indications of traffic control signaling systems at level crossings. Therefore, the behavior of pedestrians can be treated as private cars under control of a traffic control signaling system. If the related data about the pedestrians (such as the delays of urban rail-bound transport at traffic control signals and the probability of pedestrians passing through a level crossing) could be gathered, further research about the influences of pedestrians can be investigated. Similarly, the influences caused by the pedestrians in shared spaces (another kind of mixed traffic zone with urban rail-bound transport and pedestrians) can be further discussed.

In short, the behavior of private cars is used to represent all the other road traffic types for analyzing the external influences on urban rail-bound transport at level crossings in this research project. The influences of all modes of road traffic that comply with the official regulations of traffic control signaling system are studied as private cars at level crossings. For shared roads, the various speed limitations of the different road traffic types have to be taken into account for the modeling.

3.1.2   Related Findings to Road Traffic

It is always a difficult task to carry out an analysis of the behavior of mixed traffic with urban rail-bound transport and road traffic. For the simulation of road traffic, Nagel and Schreckenberg proposed to apply an one-dimensional cellular automata model for simulating freeway traffic in 1992 [Nagel & Schreckenberg 1992], which was according to the basic theory of cellular automata rules by [Wolfram 1986]. Later, Fukui and Ishibashi described an one-dimensional cellular automata model including private car movements at high speed [Fukui & Ishibashi 1996] based on the theory of the Nagel–Schreckenberg model. With the Nagel-Schreckenberg model, [Rickert et al. 1996] studied further improvements for road traffic with more realistic circumstances. Afterwards, [Blue et al. 1997] took the behavior of pedestrians into account for the modeling of large-scale open areas. For individual vehicle traffic, some other models were also investigated. [Newell 2002] constructed a new model for computer simulation to analyze the behavior of successive private cars on a homogeneous highway based on the car-following model close to real traffic under certain assumptions and constraints in [Gipps 1981].

However, all of this research only focused on the interactions between individual vehicles or pedestrians and among them for the road traffic itself, without consideration of urban rail-bound transport. The interactions between urban rail-bound transport and road traffic are barely discussed in the existing research and findings. The majority of the research is focused on the urban mixed traffic without considering the road traffic or urban rail-bound transport.

The preliminary study on the evaluation of the operating performance and operational quality of urban rail-bound transport with the influences of road traffic in mixed traffic zones using capacity research and the simulation method is discussed in this research project.

3.2 Conditions of Investigated Mixed Traffic Zone

In order to analyze how road traffic influences urban rail-bound transport, it is necessary to analyze the conditions of the investigated mixed traffic zones. For the mixed traffic zone, it may have different characteristics of topology structures, different operational regulations, as well as different road traffic loads and traffic flows of urban rail-bound transport.

The operational regulations are defined by the traffic control signaling system and are described in Subchapter 3.2.1. When the traffic loads of road traffic are very low, the influences caused by the road traffic may also exist but are generally not significant. The road traffic causes the influences on urban rail-bound transport, which can be mainly considered as the traffic loads of road traffic. Various parameters that are related to the traffic loads of road traffic were analyzed and are described in Subchapter 3.3.

3.2.1 Traffic Control Signaling System in Mixed Traffic Zones

The traffic control signaling system on urban roadways plays an important role. In mixed traffic zones, it directly determines the movements of road traffic at level crossings and indirectly on shared roads[6]. Further, it influences the operations of urban rail-bound transport.

There are various situations of traffic control signaling systems in mixed traffic zones. For different mixed traffic zones, the operational regulations of the traffic control signaling systems are also different. Traffic control signaling system facilities are usually

---

[6] The movement of road traffic on shared roads is indirectly controlled by the traffic control signals at the level crossing connected to the shared road.

installed at urban intersections. In general, there are three major possibilities for traffic control signaling systems in mixed traffic zones.

- With a traffic control signal
- Without a traffic control signal
- With a railroad crossing signal

With traffic control signaling system facilities at level crossings, the operations of the mixed traffic are controlled by traffic control signals. The traffic controller captures input data using detectors (sensors) to obtain the present state of vehicles or road users. The application of urban rail-bound transport at the traffic control signal facilities of level crossings is also conducted using different detectors along the rail tracks before the traffic control signal of the level crossing. The detector is a wireless applying point (Funkmeldepunkt) on the rail track. It is a technical device used to transfer the application to relevant signals, which is then implemented in the sketch of the applying plan [Morman 2014].

At level crossings, urban rail-bound transport operates on the basis of scheduled timetables with stochastic perturbations, as shown in Figure 3-1 (left). The headway of urban rail-bound transport is different from the time interval between successive vehicles in road traffic. When there is no urban rail-bound transport passing through the level crossing, the road traffic will be ongoing following the scheduled cycle time of the traffic control signaling system for the road traffic or self-adjusted as required. When an urban rail-bound transport is arriving, the application for traffic control signal is transferred to allow it to pass through the level crossing.

Normally, urban rail-bound transport applies its green light phase during one of the road traffic green light phases. In this situation, unscheduled waiting time of urban rail-bound transport may occur as a result of influences from the road traffic. For shared roads, if an urban rail-bound transport is hindered by road traffic, it means the shared road is occupied in front. Accordingly, the urban rail-bound transport and road traffic are indirectly controlled by the traffic control signal in front of them. In some cases, there is another technical facility located nearby the traffic control signal. Tracon (Track Circuit Controller) is an electrical device used to detect the absence of a train on rail tracks. It is designed to indicate the presence of an urban rail-bound transport vehicle in the case a failure occurs.

The operation of mixed traffic is controlled by traffic control signaling systems with the maximum red light phase duration for road traffic, as mentioned above. In this case, external influences may occur when the mixed traffic zone is a relatively complex structure with a higher traffic flow of urban rail-bound transport. The priority of road traffic rises higher than urban rail-bound transport leading to extra waiting time. Even if there are no cars, the urban rail-bound transport is theoretically still influenced by traffic control signals at level crossings. Due to the lower traffic loads of road traffic, the influences are relatively weak in the operation process.

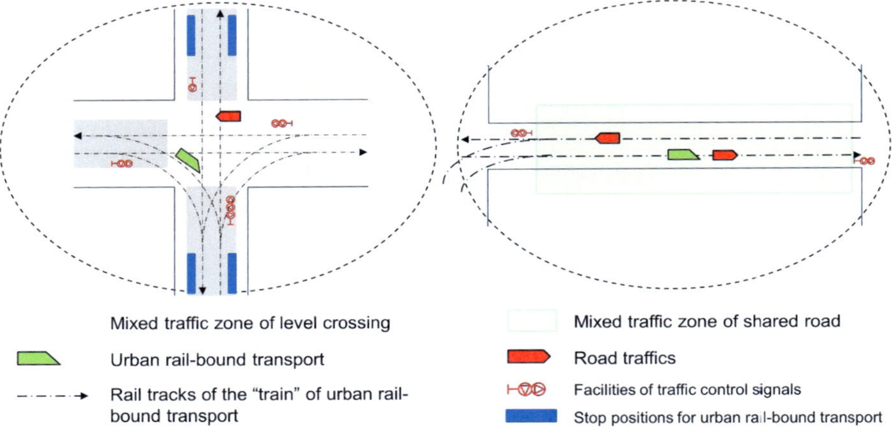

Figure 3-1: Two Conditions of Urban Rail-bound Transport with Road Traffic Influences Discussed in this Research Project

For the shared road, as shown in Figure 3-1 (right), urban rail-bound transport operations and road traffic movements share the same urban road. Along the shared road, in low traffic flow periods, there is nearly no road traffic nor traffic congestion: the influences caused by road traffic along the shared road are low. During other time periods, mostly external influences along shared road arise due to the higher traffic loads of road traffic. Urban rail-bound transport generally has a higher priority in mixed traffic zones when there is no installation of a traffic control signaling system. Such zones are relatively small and the impact on the urban rail-bound transport is consequently weak.

For the situation of a railroad crossing signal, the basic signal consists of (flashing) red (and yellow) traffic lights and a crossbuck for the road traffic. The railroad crossing signals will be activated for a time period (measured in seconds) before the train arrives. In this situation, the urban rail-bound transport has the absolute higher priority. Therefore, the influences of road traffic at mixed traffic zones without traffic control signaling systems or with railroad crossing singles are neglected in this research project.

In summary, the influences on urban rail-bound transport caused by road traffic are mainly focused on two conditions, as shown in Figure 3-1: at level crossings with a traffic control signal, and on shared roads with the hindrance of road traffic indirectly controlled by the traffic control signal at the connected level crossing.

3.2.2   Special Conditions

Certain special or unusual conditions directly affect urban rail-bound transport. These are weather; technical problems; when the road traffic is non-compliant with traffic rules, mishandling, wrong turns of private cars on the main roads; emergencies, ambulances or fire vehicles, unpredictable accidents, and special events with large amounts of people.

In this research project, weather conditions can influence the traffic loads of road traffic, which consequently influence the urban rail-bound transport. Compared with a sunny day, rainy and snowy days may affect the sight of drivers of motorized road traffic; their speed reduction indirectly increases the traffic loads of road traffic. If there is sufficient data, the impacts of road traffic on urban rail-bound transport due to weather conditions in the operation process can be analyzed in future research.

Sometimes in urban areas there are emergency situations that disturb the normal movements of urban mixed traffic. In the case of a fire or other emergency situations, fire trucks and ambulances have the absolute priority to pass through mixed traffic zones at level crossings and shared roads (except railroad crossings if a train has already reached the approaching distance), thus causing other traffic (urban rail-bound transport and road traffic) to wait longer until the situation has been cleared. Sometimes drivers or pedestrians violate the general traffic rules by taking a wrong turn or by passing through the urban roads just in front of the ongoing urban rail-bound transport. Such sorts of situations are also excluded in this research project. Because these special conditions are very random they can be discussed in future research when the appropriate data is available.

3.3 Influential Parameters

Based on the analysis of urban mixed traffic zones, road traffic influences depend mainly on the traffic loads of the road traffic. The traffic loads are affected due to different reasons such as trip purposes, means of transport, etc. These possible reasons are summarized in time-related influential parameters, which impact the waiting time of urban rail-bound transport in the investigated mixed traffic zones.

In transportation simulation research, performance usually varies at different time periods of day and on different days of the week. The time period of day and days are important influential parameters in the analysis of travel behavior and traffic modes. The time related influential parameters are mainly comprised of the time periods of day and days of operation.

Time Periods of Operation

Three time periods of operation are considered:

- HTP: High traffic flow period :7:00–9:00 and 15:00–18:00
- LTP: Low traffic flow period: 22:00–6:00
- NTP: Normal traffic flow period: the other remaining hours

Different time periods show different parts of the day with different traffic states. High traffic flow periods are called rush/peak hours, which is a part of the day with the highest traffic congestion on roads and when there is crowding on public transport. Comparatively, the other two traffic flow periods can be considered as off-peak hours. During the LTP, the road traffic influences almost can be negligible. The rest part of the day is the called normal traffic flow period, at which the traffic loads are in between the high traffic flow period and low traffic flow period.

Therefore, in the three operation time periods of the day, the traffic loads of road traffic are different. Accordingly, the road traffic influences on urban rail-bound transport are different. Hence, the time periods of operation should be considered as an influential qualitative parameter.

Day of Operation:

- Monday to Friday
- Saturday
- Sunday and holidays

Road traffic loads also differ depending on the day of the week. In time period traffic modeling, a day is sectioned into several intervals in which traffic patterns are relatively consistent. From Monday to Friday the traffic loads in urban areas are heavier than on Saturdays or on Sundays and holidays. The different travel behaviors of different days result in the corresponding delay distributions of urban rail-bound transport. Therefore in this research, considering three different types of days as an influential qualitative parameter is reasonable.

When analyzing the operations in mixed traffic zone, the time related influential parameters described above directly affect the traffic loads of road traffic, which further affect the waiting time of urban rail-bound transport.

3.4 Sensitivity Analysis

The time-related influential parameters are qualitative. In order to analyze the sensitivity of the qualitative influential parameters, it is necessary to analyze the specific conditions of each parameter. A sensitivity analysis is a way to study the uncertainty in the output of a mathematical model or system with apportioned different sources of uncertainty in its inputs [Saltelli et al. 2010]. Sensitivity analyses have been applied in the field of railway systems to test the robustness and vulnerability of a designed railway infrastructure, scheduled timetable with adjusting the corresponding parameters. In general, sensitivity analysis are applied in various fields such as environmental computer models, social sciences, and so on, which can be adapted for the sensitivity analysis with the basic methods in the simulation of operation in a rail-bound transportation system.

The sensitivity can be discerned with single or multiple changes of the values of the concerned parameters in a studied mathematical model or system. The assessed results of the importance to each parameter in the model are used to select the most effective parameters for research or project evaluation. Sensitivity analysis with modeling provides an evaluation of the confidence in the model, which estimates the contribution of quantified uncertainty in each input to the output uncertainty. The impotence, the strength and relevance of the inputs in determining the variation in the output can be ordered [Saltelli et al. 2008].

In the rail-bound simulation model, it is also important to undertake a sensitivity analysis, because it can be used to evaluate the impact of the uncertainty of various parameters on the results of the model. Especially for the qualitative influential parameters mentioned above, it is necessary to evaluate and verify the impacts of the influential parameters under different specific conditions, the external influences on urban rail-bound transport at mixed traffic zones are various. Afterwards, the significance of different specific conditions of influential parameters can be analyzed for following research.

3.4.1 Existing Methods of Sensitivity Analyses

The univariate sensitivity analysis and the multivariate sensitivity analysis are two primary methods for conducting a sensitivity analysis, which are introduced as following:

- Univariate Sensitivity Analysis (One-at-a-time)

One of the simplest forms and most common methods of sensitivity analysis is to simply change one-factor-at-a-time (OAT/ OFAT) over a range in the model by a given amount, and examine the effect that the change has on the results of model (see [Hamby 1995]).

This method of sensitivity analysis involves changing one input variable and keeping the others values fixed. Then it is measured by monitoring changes in the output (results of model). This analysis can be done repeatedly for each of the other various parameters in the same way in the model. [Saltelli et al. 2010] This approach can roughly test the sensitivity of each independent input parameter by altering each influential parameter within its domain, the output results can be estimated under the varying of them respectively and judge the significance of each of those influential parameters to the studied model or system.

In spite of the simplicity of this method, however, it is hard to fully explore the sensitivity of the input (variables), since it cannot take into account the simultaneous variation of different input variables. This means that the univariate sensitivity analysis method cannot detect the presence of interactions between various input variables [Czitrom 1999].

- Multivariate Sensitivity Analysis

In order to test different parameters simultaneously, a multivariate sensitivity analysis is carried out based on a univariate sensitivity analysis. A multivariate sensitivity analysis may be assessed by using various sampling methods, which enables the results of a deterministic model to be interpreted within a statistical framework and allows the simultaneous variation of the values of all input variables [Mckay et al. 2000] and [Blower & Dowlatabadi 1994]. However, it makes the method more complex and difficult essentially with probabilistic and algorithmic problems [Okais et al. 2010].

3.4.2 Sensitivity Analysis of Qualitative Influential Parameters

The influential parameters of the time periods and the day of operation were described in a previous chapter. The sensitivity analysis is carried out for two aspects: the variation of delays caused by road traffic along with different time periods of operation and the different days. It is necessary to analyze the road traffic influences on urban rail-bound transport. However, the data of road traffic influences is difficult to obtain. However, it is possible to gather the data of the waiting time of urban rail-bound transport at a mixed traffic zones (an investigated example of a mixed traffic zone level crossing) under the various specific conditions based on the defined influential parameters. There are two ways to collect urban road traffic flow.

- Available statistic data
- On-site measurements

Figure 3-2 shows the probability distributions of the delays (waiting times) during the three different time periods in a day. The probability distributions are based on the data obtained from on-site measurements in this example. The statistical data base is measured for two weeks with four lines of urban rail-bound transport in two directions. One hour was measured for each of the time periods. Due to a limitation of measurement conditions during the low traffic flow period, the total number of measured trains is 297. During the normal traffic flow period, the number of measured trains is 515, and the number of measured trains is 585 during the high traffic flow period.

The waiting times are each classified into 6 seconds with the corresponding number of trains, as shown in Appendix Table 1. Accordingly, Figure 3-2 can be derived to roughly show the road traffic influences on urban rail-bound transport in various time periods of operation.

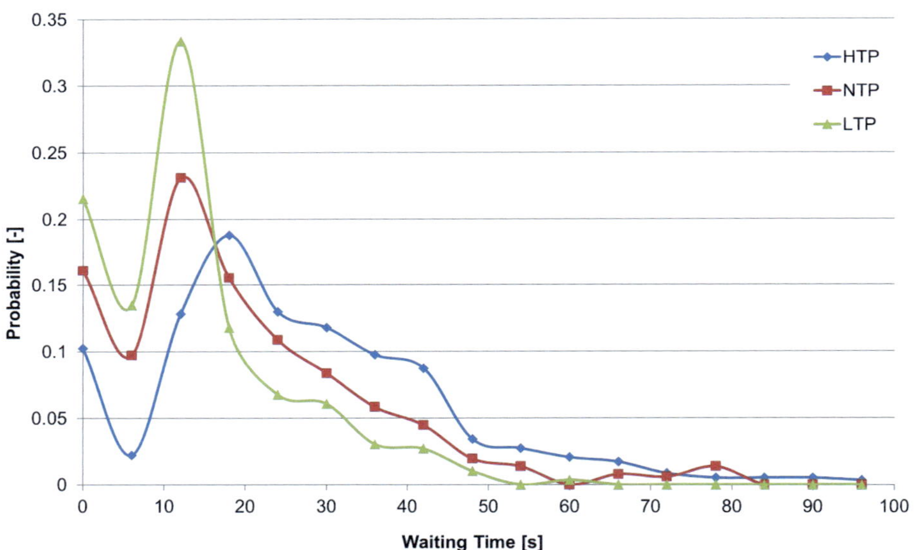

Figure 3-2: Probability Distribution of Urban Rail-bound Transport Waiting Time with Road Traffic influences at Different Investigated Time Periods (Source: modified based on [Martin & Di Liu 2016])

From the comparison of the three lines in Figure 3-2, the probability of the waiting time (delay) of urban rail-bound transport, which is statistic to zero is lower during the high traffic flow period. Delays that are zero mean that the urban rail-bound transport is punctual or earlier than the scheduled time. In the low traffic flow period, there is higher probability to be punctual based on the scheduled timetable. According to the fluctuation of the three lines, the probability of occurrence of very small waiting times (about 6 seconds) is relative low.

The waiting time that is most likely to occur is when the delays are also relatively small (about 12 seconds), the probability has the same trend for all three of the time periods, which is distinctly higher during low traffic flow period. It is reasonable that the probability of smaller waiting times occurring is higher; on the contrary, the probability of larger waiting times is lower during the low traffic flow period.

The probability of large delays (more than 18 seconds) occurring during the low traffic flow period decreases rapidly and is distinctly lower than the other time periods. Regardless if the delays are small or large, the distribution probability of normal traffic flow period is in between. So in the process of capacity research with consideration of urban mixed traffic, three characteristic time periods of operation should be distinguished for analyzing the influences of the different traffic loads.

For the different days of operation, about 2.5 weeks of data was collected. For weekdays (Monday to Friday), 10 days of measurement were taken: 1,023 trains were counted. For Saturday, 4 days of measurements were taken: 370 trains were counted. For Sunday and holidays, 4 days of measurements were also taken: 371 trains were counted.

These measured waiting times are also each classified into 6 seconds with the corresponding number of trains (see Appendix Table 2). The probability of occurrence of each class of waiting time during the different days can be shown in Figure 3-3.

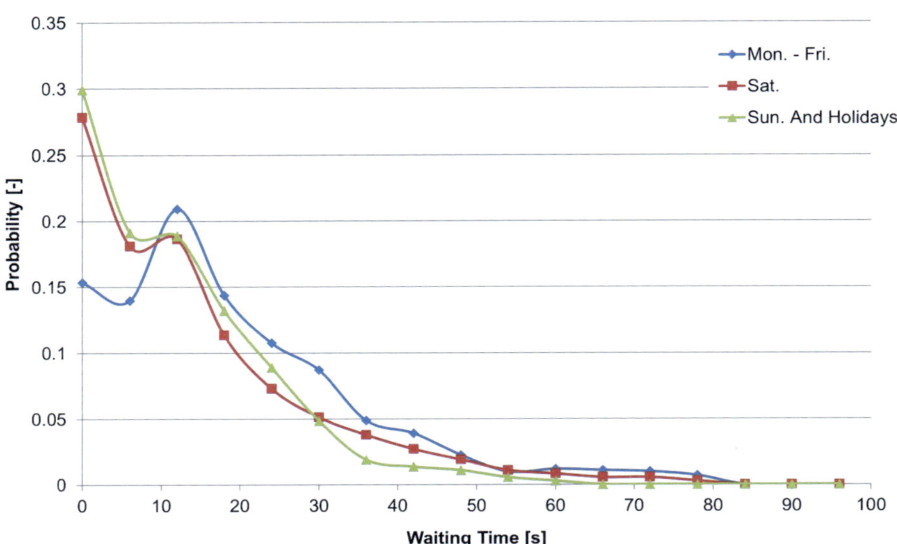

Figure 3-3: Probability Distribution of Urban Rail-bound Transport Waiting time with Road Traffic Influences on Different Day Types (Source: modified based on [Martin & Di Liu 2016])

In Figure 3-3 it can be seen that the probability of the waiting times (delays) that are statistically close to zero is lower on weekdays (Monday to Friday). It is slightly higher on Sunday and holidays than on Saturday. When there is 0 probability of delays, this means that the urban rail-bound transport is punctual or early. The probability of punctual operation is highest on Sunday and holidays. For the weekdays, it is obviously lower than others, which means the punctuality rate is lower.

The probability of occurrence of very small waiting times (about 6 seconds) goes down from the probability of punctual, which means there is a lower chance that they will occur. When the delays are relatively larger (more than about 10 seconds), the probability of delays on Monday to Friday increases and tends to be the highest of the other days. It is also reasonable that the longer waiting times have a higher probability of occurrence on weekdays.

In this case, the probability of delays about 30 seconds on Saturday is a little smaller than that on Sunday and holidays. This trend reverses when the delays exceed over 30 seconds. It means that the occurrence of larger waiting times on Saturday has a relative higher probability than on Sunday and holidays. However, the probability distributions of delays on Saturday as well as on Sunday and holidays have no significant differences from the statistical point of view. Therefore, in the process of capacity research at mixed traffic zones, it is meaningful only to distinguish the day types into two categories:

- Monday to Friday (weekdays)
- Saturday, Sunday and holidays

4 Modeling Approach for Road Traffic Influences on Urban Rail-Bound Transport

For urban mixed transportation, in addition to the multifarious processes of urban rail-bound transport operation, road traffic is operated under the control of traffic control signals and further influences urban rail-bound transport operations in mixed traffic zones, as described in Chapter 3.

These road traffic influences have an impact on the simulation results of capacity research on urban rail-bound transport, which leads to a deviation of the existing waiting time function. In order to systematically describe the various road traffic influences, two approaches were developed in this research project. In this chapter, the modeling approach is depicted to build an integrated simulation model.

4.1 Basic Concept

The modeling approach considers road traffic as urban rail-bound transport in the model of railway operation with the help of the synchronous simulation tool RailSys [RMCon 2010] and assistant software PULEIV [Martin et al. 2011]. For road traffic, some other simulation tools such as VISSIM[7] can also be used. However, these tools do not fulfill the requirements of this research, because the operations and interactions of urban rail-bound transport (e.g. train movements with consideration of railway signaling systems) as well as the operating program cannot be also modeled exactly related to capacity of urban ail-bound transport.

The term "modeled road traffic" is used to refer to the road traffic under the control of traffic control signals in the simulation tool. The term "train" is used to refer to the modeled trains in the simulation model representing the urban rail-bound transport movements. The road traffic (private cars) at level crossings are modeled as urban rail-bound transport directly in the simulation model [Martin et al. 2012]. Through the interaction between the "trains" and "modeled road traffic" in the simulation process,

---

[7] VISSIM: **V**erkehr **I**n **S**tädten – **SIM**ulationsmodell, which can be used for simulation of multi-modal system and public transport in microscopic scale developed by PTV Planung Transport Verkehr AG in Karlsruhe, Germany. Fellendorf & Vortisch [2010]

the influences of road traffic on urban rail-bound transport can be observed and investigated.

4.2 Methodology for Modeling Approach

4.2.1 Simulation Model

For this approach, the most important task is to build the simulation model so that the road traffic influences can be reflected to a sufficient extent. To integrate the trains of "modeled road traffic" in the simulation tool, two important inputs should be prepared:

- Topology structure of the investigated mixed traffic zone
- Representative estimation of the "scheduled timetable" for the "modeled road traffic"

The topology structure describes the interaction characteristics between the "trains" and the "modeled road traffic" in the form of the layout of the investigated mixed traffic zone. According to the topology structure, the infrastructures[8] can be built in the simulation tool for both the "trains" (urban rail-bound transport) and the "modeled road traffic" (road traffic). The rail-bound infrastructure for "trains" of urban rail-bound transport in the simulation tool RailSys is built for railway operation on the investigated rail-bound infrastructure (with rail tracks, switches, signaling systems, stations, etc.) as close to reality as necessary. The infrastructure for "modeled road traffic" has to be built in the simulation tool too, which is based on the layout of the given road lanes with their directions of traffic flow at the investigated mixed traffic zone.

At level crossings, according to analysis of the road lanes that interfere with rail tracks for urban rail-bound transport, the corresponding directions of the road traffic are also included in the model. Each road lane that interferes with rail tracks is built as a new track line with corresponding modeled stations for departure and arrival in the simulation model. The movements of the road traffic are controlled by traffic control signals at the level crossings, therefore, the traffic loads of the road traffic that

---

[8] The infrastructures should include the rail-bound infrastructure for "trains" of urban rail-bound transport, and the urban roadways that are also modeled as rail-bound infrastructure for the trains of "modeled road traffic".

may affect the time duration of occupying the level crossing can be represented by the corresponding traffic light phase of the traffic control signals.

According to the regulations of the red light phase limitation, the train of the "modeled road traffic" must have the right to occupy the block of the level crossing when it has already waited for the time duration of red light phase limitation. The time duration of the red light phase limitation can be seen as the headway of the trains of the "modeled road traffic" at the level crossing. This headway of the trains of "modeled road traffic" can be made more accurate by adjusting the model with data from on-site observation. In addition, the corresponding blocking time of the trains of the "modeled road traffic" is set as the occupation time with its minimum green light phase. It can be also adjusted more accurately based on the data collected from the on-site observations. In summary, the basic timetable for the trains of the "modeled road traffic" can be scheduled for the simulation model in the simulation tool RailSys.

However, for the mixed traffic zones of shared roads, the rail tracks for urban rail-bound transport are also utilized for the trains of the "modeled road traffic". The operation of urban rail-bound transport is not only dependent on the movements of the road traffic controlled by traffic control signals, but also dependent on the various traffic modes and the traffic loads of the road traffic. It is hard to model the interactions between urban rail-bound transport and road traffic with the same concept. Therefore with this modeling approach, shared roads are considered as a series of fictitious connected level crossings.

When the urban rail-bound transport runs along the shared road, it can be hindered by road traffic at any point. When the road traffic in front of the urban rail-bound transport starts its green light phase, the urban rail-bound transport can also run for the same time duration. Therefore, the trains of the "modeled road traffic" are positioned at the fictitious level crossings that interfere with the urban rail-bound transport along the shared road. The "modeled road traffic" has headway similar to the green light phase duration, and the blocking time is equal to the non-green time phases of the road traffic controlled by the traffic control signal at the next adjacent level crossing.

The distance between two fictitious level crossings is determined as the distance that the urban rail-bound transport runs for the occupation time of the road traffic controlled by the traffic control signal. Due to the large amount of time of effort to model the many level crossings, there are very high costs to evaluate the shared road with this modeling approach, nevertheless this method can be used for the mixed traffic zone of level crossing.

The infrastructure for an investigated example of mixed traffic zone of a level crossing is shown in Figure 4-1. There are four operation lines ($L_{URT,1}$ - $L_{URT,4}$ see Appendix Table 3) with double rail tracks for "trains" of urban rail-bound transport, and four operation lines ($L_{ROT,1}$ - $L_{ROT,4}$) with single rail tracks for the trains of "modeled road traffic" at the level crossing. There are eight stop positions (in front of the signals and exit points) at the level crossing in the simulation model with directions ($S_{W,ROT/URT}$, $S_{E,ROT/URT}$, $S_{N,ROT/URT}$ and $S_{S,ROT/URT}$ [9]) for road traffic and urban rail-bound transport respectively.

For example, the trains of "modeled road traffic" run on the rail track of operation line ($L_{ROT,1}$) from the entry stop $S_{E,ROT}$ to the exit stop $S_{S,ROT}$, which should interfere with six operation lines of the three double rail tracks for "trains" of urban rail-bound transport ($L_{URT,1}$, $L_{URT,2}$ and ,$L_{URT,3}$ with two routes [10] in corresponding direction respectively).

---

[9] The stations $S_{W,ROT/URT}$, $S_{E,ROT/URT}$, $S_{N,ROT/URT}$ and $S_{S,ROT/URT}$ here refer to the stations in the directions of west, east, south and north respectively for both road traffic and urban rail-bound transport.

[10] Route of mixed traffic (urban rail-bound transport and road traffic) is detailed described in subchapter 6.4.2.

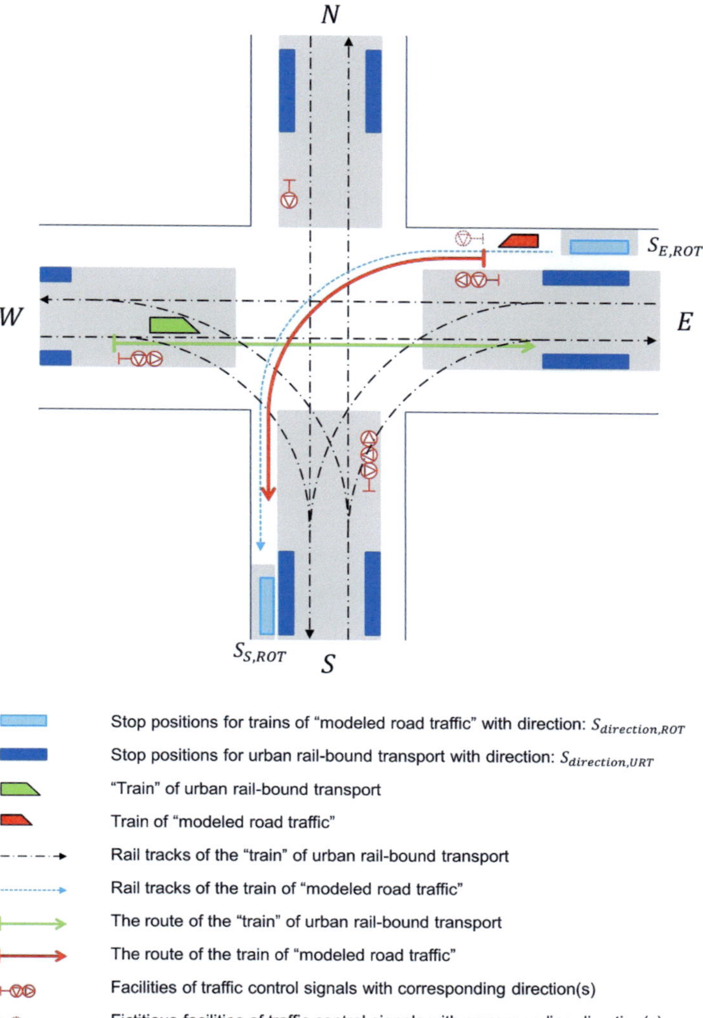

Figure 4-1: Infrastructure of an Investigated Example of Mixed Traffic Zone of Level Crossing for the Simulation Model

For this simple case, it is necessary to add one new rail track for the trains of "modeled road traffic" with two stations in the simulation model based on the built infrastructure of urban rail-bound transport. The scheduled timetable for the trains of "modeled road traffic" as the method described previous in this chapter with the adjusted headway based on the limitation of red light phase (120 seconds in this example) and the adjusted blocking time based on the practical on-site measurement is partly given as example in Table 4-1.

Operation Lines	Route	Direction	Departure Time	Arrival Time	Headway	Departure Time of Next Trip	Arrival Time of Next Trip
$L_{URT,1}$	$r_{URT,1}$	$S_{W,URT} \rightarrow S_{E,URT}$	08:03:24	08:04:38		08:13:24	08:14:38
	$r_{URT,2}$	$S_{E,URT} \rightarrow S_{W,URT}$	08:05:44	08:06:32		08:15:44	08:16:32
$L_{URT,2}$	$r_{URT,3}$	$S_{N,URT} \rightarrow S_{S,URT}$	08:04:49	08:05:22	00:10:00	08:14:49	08:15:22
	$r_{URT,4}$	$S_{S,URT} \rightarrow S_{N,URT}$	08:07:48	08:08:24		08:17:48	08:18:24
$L_{URT,3}$	$r_{URT,5}$	$S_{W,URT} \rightarrow S_{S,URT}$	08:08:24	08:09:49		08:18:24	08:19:49
	$r_{URT,6}$	$S_{S,URT} \rightarrow S_{W,URT}$	08:08:16	08:09:05		08:18:16	08:19:05
$L_{ROT,1}$	$r_{ROT,1}$	$S_{E,ROT} \rightarrow S_{S,ROT}$	08:00:21	08:00:46	00:02:00	08:02:21	08:02:46

Table 4-1: Partly Scheduled Timetable of an Investigated Example of Mixed Traffic Zone of Level Crossing for Simulation Model

When a train of "modeled road traffic" comes into and occupies this investigated level crossing, the trains of urban rail-bound transport cannot pass through at the same time. This shows that the influences of road traffic on urban rail-bound transport will occur due to the operational hindrances between trains (urban rail-bound transport and road traffic) during operation.

In the scheduled timetable, due to the addition of the trains of the "modeled road traffic", it is possible that some conflicts cannot be solved. However, it is better to try to reduce the possible conflicts between the trains (both urban rail-bound transport and road traffic) by adjusting the "trains" within a certain range (usually adjusting the headway of trains of "modeled road traffic"), so that the scheduled timetable can be made more plausible. Accordingly, some of the trains of the "modeled road traffic" in the scheduled timetable do not have a fixed headway.

Furthermore, in order to simulate the road traffic influences more reasonably, the respective priority of "trains" of urban rail-bound transport and trains of "modeled road traffic" have to be set in the simulation model. Their priority can also be used to adjust the operation of trains in the simulation process, especially when delays occur or road traffic is congested. In order to simulate the interaction between them, the "trains" of urban rail-bound transport principally have a higher priority than the trains of "modeled road traffic" in a certain range of delay (in seconds) as the basis setting for the simulation model. When the delays of the trains of "modeled road traffic" in the operation process exceed the certain defined range of delays, the trains of "modeled road traffic" get higher priority in operation. Until the delays of trains of "modeled road traffic" go back to the defined range, their priorities will come back to basic settings. When the "trains" of urban rail-bound transport are simultaneous delayed more than a corresponding defined limit, both the trains of urban rail-bound transport and "modeled road traffic" will be assigned the same level of priority.

In summary, as the basic prerequisite for capacity research, it is necessary to build the rail-bound infrastructures (for both urban rail-bound transport and road traffic) and corresponding scheduled timetables in the simulation model. Moreover, the settings of priorities for both trains of urban rail-bound transport and "modeled road traffic" also play an important role for their operations and to reflect their interactions in the simulation process as close to reality as possible.

4.2.2 Capacity Research of Modeling Approach

Capacity research for urban rail-bound transport can be carried out with the simulation method to evaluate the operating performance of the investigated area with a given operating program. It is used to determine the throughput capacity, the waiting time function as an important intermediate result, and further to derive the recommended area of traffic flow.

For pure railway systems (with no road traffic influences), the capacity research can first be carried out by establishing the infrastructure and scheduled timetable of the investigated scenario in the simulation tool (e.g. RailSys). A series of stochastic timetables with stepwise-varied traffic flows are then created randomly based on the given operating program with a selected timetable frame with the help of the software PULEIV. The simulation results are used to determine the throughput capacity and fit the curve of the waiting time function with discrete waiting time data points, and then to derive the recommended area of traffic flow by simulating the stochastic timetables with stepwise-varied traffic flows.

For capacity research of urban mixed traffic, the emphasis is on the capacity of the urban rail-bound transport on investigated area with a given operating program. Therefore, the road traffic influences on urban rail-bound transport have to be considered in the simulation model. With the modeling approach, the road traffic influences can be directly reflected in the simulation process, because the trains of "modeled road traffic" are already included in the simulation model.

The timetable frame is normally selected to be 2 hours at an investigated time period (often during the HTP) as structure of operating program. Since the road traffic is investigated as external influences of urban rail-bound transport, the traffic loads of the "modeled road traffic" will be fixed as a constant value during the investigated time period. A series of stochastic timetables with stepwise-varied traffic flows of urban rail-bound transport and fixed traffic loads of modeled road traffic are created. Based on the simulation protocols of the generated timetables through single simulations of stochastic timetables, the results of the capacity research, throughput capacity can be derived and the curve of the approximated waiting time function (described in Chapter 6) can be fitted with road traffic influences. Afterwards the recommended area of traffic flow can be derived.

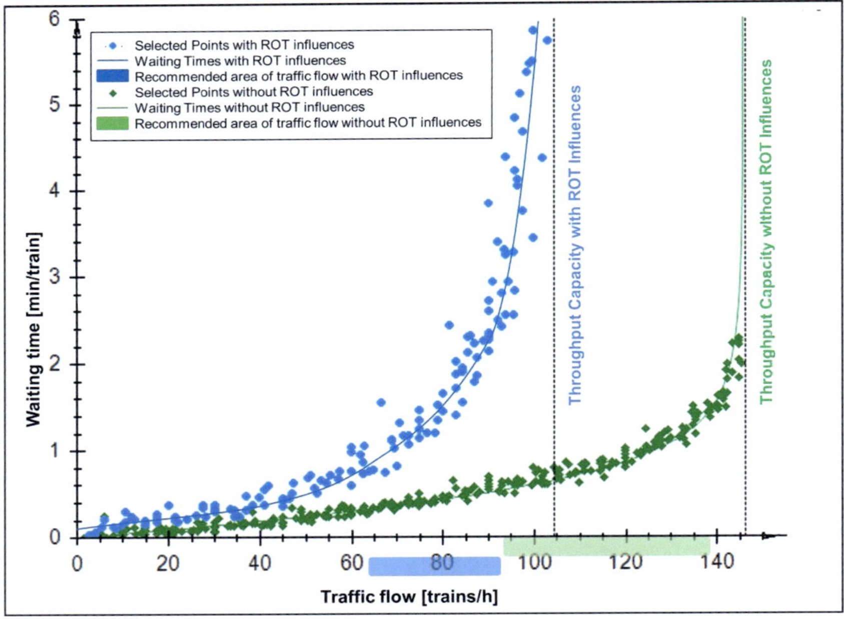

Figure 4-2: Results of Modeling Approach Comparison to that without Road Traffic Influences for an investigated Example (Source: modified based on [Martin & Di Liu 2016])

Figure 4-2 shows the results of the capacity research with the modeling approach for the investigated example of a large-scale, rail-bound network where the level crossing with the road traffic is described in Subchapter 4.2.1 (see Appendix Figure 1 and Appendix Table 3). As shown, the throughput capacity and the recommended area of traffic flow are different for the blue fitted curve with the discrete points through the developed waiting time function with road traffic influences (see Subchapter 6.3) and the green fitted curve without road traffic influences in a pure railway system.

The throughput capacity is the "maximum" traffic flow of urban rail-bound transport in a given investigated area and occurs when the waiting time approaches infinity. The blue fitted curve with the throughput capacity of 104.3 trains/h has an obvious loss compared to the green fitted curve with the throughput capacity of 146.3 trains/h because of the competition of the road traffic for occupation of the investigated level crossing, which restricts the possible capacity of urban rail-bound transport in the investigated area. The gap between the two fitted curves shows a relative increase of waiting time of urban rail-bound transport with the same traffic flow caused by the road traffic influences. Accordingly, the recommended area of traffic flow with road traffic influences (highlighted in blue) is between 60.3 trains/h and 91.7 trains/h, which is lower than that without road traffic influences (highlighted in green) between 93.9 and 139.8 trains/h.

4.3 Applications of Modeling Approach

There are two types of mixed traffic zones that have already been discussed in Subchapter 3.2: level crossings and shared roads. In the case of level crossing, the topology structure is relative clear, and the trains of "modeled road traffic" can be directly modeled with general train operations controlled by traffic control signaling system. However, in the case of shared roads, the road traffic affects the behavior of urban rail-bound transport along the whole course of the rail tracks. The dynamics of the vehicle movements between urban mixed traffic are different and the interactions between them are too flexible to model as typical real level crossings. As described in Subchapter 4.2.1, a lot of effort is required to model a shared road as a series of adjacent level crossings.

For a relative simple and small-scale mixed traffic zone, the modeling approach has the advantages that data can be collected simply to build the simulation model through modeling the road traffic as trains of "modeled road traffic". However, for large-scale, mixed traffic zones, especially in the case where shared roads are involved, the cost of data collection and building the simulation model are relative high and not as accurate. In order to overcome the disadvantages, another approach is developed, and is described in Chapter 5.

5 Distribution Approach for the Description of Road Traffic Influences on Urban Rail-bound Transport

In order to systematically describe the road traffic influences on urban rail-bound transport in urban mixed traffic zones, an approach is developed, which is used to supplement the disadvantages of modeling approach as described in Chapter 4. With this approach, the road traffic influences are considered as perturbations in the simulation model. This approach is referred to as the "distribution approach" in this research project, as not to confuse it with the described "modeling approach" in Chapter 4. It uses applicable mathematical distributions to model the road traffic influences on urban rail-bound transport, which are considered as perturbations acting on the simulation process.

5.1 Basic Concept

The stochastic influences from road traffic can be modeled as perturbations in the simulation model used here. When the perturbations are introduced into the simulation tool, the impacts of the external influences can be analyzed in the form of delays (unscheduled waiting times of trains). The perturbations are described in probability density functions for a certain probability distribution with perturbation parameters[11]. The approach is therefore called the distribution approach.

Multiple simulations are carried out to study the perturbed rail-bound system, which is disturbed by stochastic influences in the process of the operational simulation. The stochastic influences in reality can be gathered from real operations to be modeled as perturbations in the simulation. It can be an empirical distribution of actual data or the known theoretical probability distribution. In [Yuan 2006], the lognormal distribution and the Weibull distribution were proposed to model the stochastic events during railway operations. In [Wakob 1985], the inter-arrival times and the service times were assumed as a Erlang-k distribution. The negative exponential distribution and the empirical distribution from actual collected data are used in the simulation tool RailSys [RMCon 2010].

---

[11] Each perturbation is described by three perturbation parameters in the form of probability density function. The three parameters are introduced in subchapter 5.2.1.

In this research project, the multiple simulations are carried out with RailSys. The delays caused by perturbations in the simulation tool are generated by the inverse function of the cumulative distribution function based on the negative exponential distribution. Additionally, the method developed in this chapter can also be applied for other probability distributions.

Road traffic influences on urban rail-bound transport at mixed traffic zones can be considered as one kind of (external) perturbations during operations. Road traffic influences are highly related to the different operational situations. To derive suitable descriptions for the various perturbations on urban rail-bound transport, a systematic categorization of the perturbations is necessary. The perturbation parameters can be determined using data collected from available statistics or measured on-site) at stops of urban rail-bound transport in front and in back of the investigated mixed traffic zone (see Subchapter 5.2.1).

The calibration algorithm (see Subchapter 5.2.2) will be applied to derive the road traffic influences on urban rail-bound transport; it is used through comparing the simulated instantaneous value and the target value of indicators[12] in reality An algorithm to automatically calibrate the perturbation parameters has already been developed for rail-bound systems [Cui et al. 2014]. The different types of perturbations are defined as initial delay, running time extension, and dwell time extension. The perturbation parameters for each perturbation are continuously adjusted (calibrated) in order to reach the best fit between the simulated delays in using the simulation tool and the delays collected in real operations.

After the perturbation parameters of the various perturbation types are determined, the delays caused by hindrances between urban rail-bound transports themselves and/or road traffic influences in the simulation model can be derived. For the capacity research of urban rail-bound transport with consideration of road traffic influences, the perturbed single simulations will be carried out based on the setting of perturbation parameters representing road traffic influences with a given operating program in the simulation tool (see Subchapter 5.4).

---

[12] Instantaneous value and the target value of indicators are discussed also in Subchapter 5.2.2.

5.2 Perturbation Parameters for Mixed Traffic Zones

The distribution approach is based on the concept of the calibration algorithm (see Subchapter 5.2.2) and is used to determine the corresponding perturbation parameters of each perturbation (see Subchapter 5.2.1) for road traffic influences in the simulation model. With the calibrated perturbation parameters in the simulation model, the output delays of the investigated area can be produced through introducing perturbations into the simulation model.

5.2.1 Definition of Perturbation Parameters

The perturbation parameters will be defined for different perturbation types, which may result in various delays to the scheduled timetable. The stochastic road traffic influences that lead to delays in urban rail-bound transport operations can be modeled in the simulation model by way of corresponding perturbation types.

The output delays resulted from multiple simulations with perturbations are the deviations from the scheduled timetable, which will be compared to the delays in the actual operation. In this research project with consideration of urban mixed traffic, the perturbations including road traffic influences on urban rail-bound transport in operation are categorized as the following perturbation types:

- Initial delay (entry delay): the delays trains have coming into the investigated area due to perturbations occurring outside the investigated area. The initial delays are introduced at the first station of the whole investigated area in the simulation model.

- Dwell time extension: the dwell time extension describes the unscheduled time extension of dwell time for the process of passenger boarding and alighting at scheduled stations.

- Running time extension: the unscheduled time extension of running time during train-running. The delays of urban rail-bound transport are caused after completed after the boarding and alighting of passengers and before arriving at the next scheduled station.

The running time extension perturbation can be further categorized into two parts based on its causes. One is the running time extension with the perturbation parameters P_{URT} [13] due to urban rail-bound transport itself, which might be caused by various technical, human, or natural events. This occurrence of running time extension of urban rail-bound transport doesn't depend on road traffic.

The other type of running time extension with the perturbation parameters P_{ROT} [14] represents the external influences caused by road traffic at urban mixed traffic zones.

For each perturbation type, there are three perturbation parameters: del_m, $pro.$ and del_{max}, to model the theoretical distribution:

- del_m [minute]: The average value of delays.
- $pro.$ [%]: The proportion of perturbed urban rail-bound transport (delayed urban rail-bound transport).
- del_{max} [minute]: Maximum delay for perturbed urban rail-bound transport.

These three perturbation parameters can be used to describe the probability distribution and the probability density function of each perturbation in the simulation model. Two parameters (del_m and $pro.$) are considered for the calibration as mentioned above. By using the calibration process, the values of these perturbation parameters are iteratively adjusted to minimize the difference between the output delays (instantaneous values of indicators) resulted from multiple simulations and the actual delays (target values of indicators) in operation.

5.2.2   Basic Calibration Algorithm

The goal of calibration in this research project is to determine and calibrate the values of the defined perturbation parameters (see Subchapter 5.2.1). By comparing the given/collected target values of the indicators (i.e. actual delays in the investigated area) and the instantaneous values of the indicators (i.e. the derived delays resulting from the output of multiple simulations), the parameters are calibrated in an iterative calibration process.

---

[13] P_{URT} here indicates three perturbation parameters for running time extension due to urban rail-bound transport itself.

[14] P_{ROT} here indicates three perturbation parameters for running time extension due to external influences caused by road traffic.

The calibration algorithm is carried out using two indicators with normalized values: the average value of delays (in minutes) and the proportion of delayed trains, i.e. urban rail-bound transport (in %).

The target values of the two indicators are given from the collected operational data in real operation. The instantaneous values of the two indicators are obtained from multiple simulations with the inputs of the current values of the perturbation parameters. When an urban rail-bound transport train arrives earlier than the scheduled arrival time, the value of the delay will be set as zero.

The calibration process is an iterative process, which includes several rounds of multiple simulations. After each round, the simulation results (instantaneous values of two indicators) will be compared to the reality (target values of two indicators), and the values of the perturbation parameters will be calibrated. In the calibration process, the current values of the perturbation parameters will be given as input for each round of calibration. The value of perturbation parameters will be adjusted iteratively through comparison of the given target values and the instantaneous values of two indicators in the calibration process. The difference of the target value and the instantaneous value will be reduced continuously until they converge.

Therefore, the error of each iterative round is defined as the difference between the target values and the instantaneous values of the two indicators. The instantaneous values are resulted from multiple simulations with current perturbation parameters in this iterative round. The errors can be calculated and further minimized with adjustment of the perturbation parameters in an iterative calibration process based on machine learning theory, which is defined in [Cui et al. 2014]. The iterative process of calibration can be regarded as an optimization process. The values of the calibrated perturbation parameters can be derived to generate the perturbations of the road traffic influences.

5.3   Methodology of the Distribution Approach

5.3.1   Overview

In this research project, a method is developed to model the influences from road traffic with the distribution approach. In order to carry out capacity research for urban rail-bound transport with consideration of mixed traffic, the road traffic influences can be regarded as perturbations in the simulation model. Based on the defined perturbation types (initial delay, dwell time extension, and running time extensions) and the corresponding perturbation parameters of a perturbation in the simulation model, multiple simulations can be carried out for capacity research.

The perturbations caused by road traffic are modeled as additional running time extensions with the perturbation parameters P_{ROT}, which can be used for capacity research to determine the throughput capacity and fit the waiting time function. Furthermore, the recommended area of traffic flow with consideration of road traffic influences will be derived. In order to calibrate the perturbation parameters for the running time extension P_{ROT} with multiple simulations, the values of the other perturbation parameters with various types of perturbations should be inputted into the simulation model.

For an investigated mixed traffic zone, the perturbation parameters of initial delays at the first station of each train run can be obtained from actual collected data and inputted into the simulation model in the form of an empirical distribution or a theoretical statistic distribution. Similarly, the perturbation parameters of dwell time extensions at each scheduled station can also be obtained from actual collected data at each scheduled station. In addition to the running time extension modeled by the perturbation parameters, P_{URT} is the running time extension of urban rail-bound transport between two stations without consideration of external influences caused by road traffic. The perturbations parameters P_{URT} are applied in the simulation tool to model and generate stochastic influences, which are only caused by urban rail-bound transport.

In this approach, the running time extension with the perturbation parameters P_{URT} will be the same at any time period of an operation day. The various technical, human, and natural events are assumed to be independent of a specific time period or a location of operation. These events can be causes for the occurrence of a failure in the railway system. The human reasons for running time extensions in this research are regarded as one general factor for any operation environment. The influences of humans are modeled in the running time extension with the perturbation parameters P_{URT}, for any time period and location of operation.

Unlike with pure railways, there are also perturbations caused by road traffic that have to be considered when dealing with a mixed-traffic zone. The perturbations caused by road traffic influences are regarded as the running time extension for the urban rail-bound transport with the perturbation parameters P_{ROT}. The waiting time for the urban rail-bound transport caused by the road traffic is one of the important outputs of the model.

The delays in the case without road traffic are hard to be collected in real operation, since it is impossible to collect without all road traffic for data collection. Instead, the delays during the low traffic flow period (22:00–06:00) are used with the assumption that the influences of the road traffic on urban rail-bound transport at the low traffic flow period can be ignored. In reality, the reduction of velocity for urban rail-bound transport is reasonable even during the low traffic flow period for safety reasons. It is assumed that no road traffic influences exist to simplify the model. Furthermore, since the priority of urban rail-bound transport is high and the density of urban rail-bound transport is low, almost no knock-on delays caused by road traffic arise in mixed traffic zones during this time period.

Therefore at low traffic flow periods, there is only one perturbation of running time extension with the perturbation parameters P_{URT}. The delays of urban rail-bound transport in real operation (target values of indicators) are just the delays caused by urban rail-bound transport itself due to the operational hindrances between trains, which can be obtained from the actual collected data. The running time extension caused by road traffic influences can be considered as zero. Therefore, the running time extension caused by urban rail-bound transport itself can be derived from the calibration process during the low traffic flow period based on actual collected data of initial delays and dwell time extension in the form of an empirical distribution. Furthermore, with the derived running time extension with the perturbation ter P_{URT}, the running time extension caused by road traffic influences with the perturbation parameters P_{ROT} can also be determined with the iterative calibration process presented in Subchapter 5.3.2.

5.3.2   Procedure of the Distribution Approach

The procedure of the distribution approach is an iterative process. Two steps to determine the running time extension with the perturbation parameters P_{URT} (without influences of road traffic) and P_{ROT} (with influences of road traffic) are carried out iteratively in the simulation model:

- Determination of the running time extension with the perturbation parameters P_{URT}: the influences of road traffic during low traffic flow period are not considered.
- Determination of the running time extension with the perturbation parameters P_{ROT}: the influences caused by road traffic during the investigated time periods are included.

A calibration process for urban rail-bound transport without any influences of road traffic is firstly carried out during a low traffic flow period in order to derive the perturbation parameters P_{URT} for the running time extensions. The running time extension with the parameters P_{ROT} caused by road traffic is further calibrated until the simulated delays (instantaneous values of the indicators) caused by the urban rail-bound transport and road traffic can fit the delays collected in real operations (target values of the indicators) during the investigated time periods.

As the basis of the distribution approach, the correlated statistical data of the initial delays at the first scheduled station of each train run in the investigated mixed traffic zone, as well as the dwell time extensions at each scheduled station are necessary. The infrastructure and the timetable used to simulate the train runs of urban rail-bound transport are the input data of the simulation model. The corresponding statistical data of the surrounding stations should be obtained to determine the running time extension caused by road traffic for the investigated mixed traffic zone. The calibration process is shown in Figure 5-1.

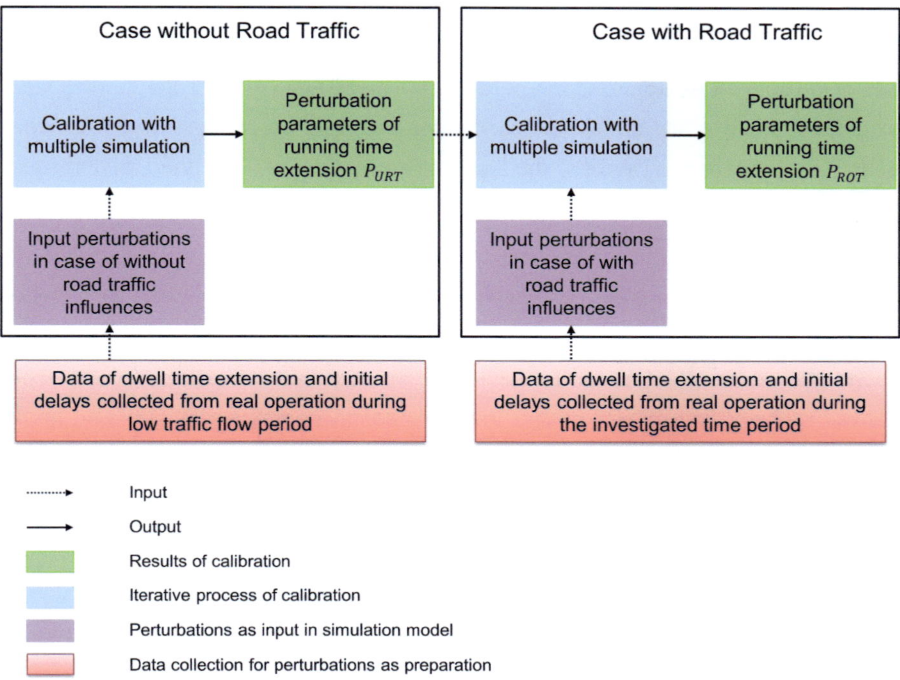

Figure 5-1: Procedure for Determination of the Perturbation Parameters P_{ROT} (Source: modified based on [Martin & Di Liu 2016])

According to the collected statistical data of the initial delays and the dwell time extensions of the scheduled stations nearby the investigated mixed traffic zone at low traffic flow period, the two perturbation types can be introduced with the corresponding perturbation parameters in the simulation model with empirical distributions or theoretical statistic distribution (e.g. negative exponential distributions in RailSys). The initial values of the running time extension without the road traffic influences with the perturbation parameters P_{URT} are inputted into the simulation model. The multiple simulations can be carried out with the scheduled timetable of a low traffic flow period. The initial delays and the dwell time extensions with the perturbation parameters are given from actual statistical data. The running time extensions are generated from the given initial values of the perturbation parameters P_{URT} on the investigated infrastructure.

Through the calibration process, the simulation results of the output delays (instantaneous values of indicators) are iteratively compared to the actual delays (target values of indicators). By adjusting the perturbation parameters P_{URT} for the running time extension during a low traffic flow period, the error will be minimized until convergence is reached. Hence, the approximate values of the perturbation ters P_{URT} for the running time extension without road traffic influences can be determined.

During the investigated time period with high traffic flow, the delays caused by road traffic influences have to be considered. The running time extension with the perturbation parameters P_{ROT} has to be taken into account in the simulation model. The derived running time extension with the perturbation parameters P_{URT} is also introduced into the simulation model as perturbations in the investigated time period. The corresponding initial delays and dwell time extensions of each scheduled station nearby the investigated mixed traffic zone can be also introduced into the simulation model as perturbations with the perturbation parameters collected from the statistics of available data or data measured on-site during the investigated time periods.

In the iterative calibration process, the values of the perturbation parameters P_{ROT} for the running time extension in the investigated time period are adjusted iteratively until the error is minimized and the values converge. The approximate values of the perturbation parameters P_{ROT} for the running time extension caused by the road traffic influences can be determined for capacity research on urban rail-bound transport.

5.4 Capacity Research with the Distribution Approach

For capacity research of urban rail-bound transport, the road traffic influences have to be considered in the simulation model as perturbations. After the perturbation parameters are determined and calibrated, random perturbations caused by road traffic can be introduced into the operational simulation. The resulted waiting times are used to carry out the capacity research for mixed traffic zones.

The situation without the influences of road traffic will be investigated with the determined perturbation parameters P_{ROT} of the running time extension. In the step of creating a series of stochastic timetables with stepwise-varied traffic flows, the timetable without road traffic influences will be established. The randomly-created timetables can be regarded to a certain extent as the timetables with perturbations of initial delays. For the investigated mixed traffic zone, the perturbation of the running time extension along with the derived the perturbation parameters P_{ROT} for road traffic influences will be considered. The perturbations will be generated and introduced into these randomly-created timetables.

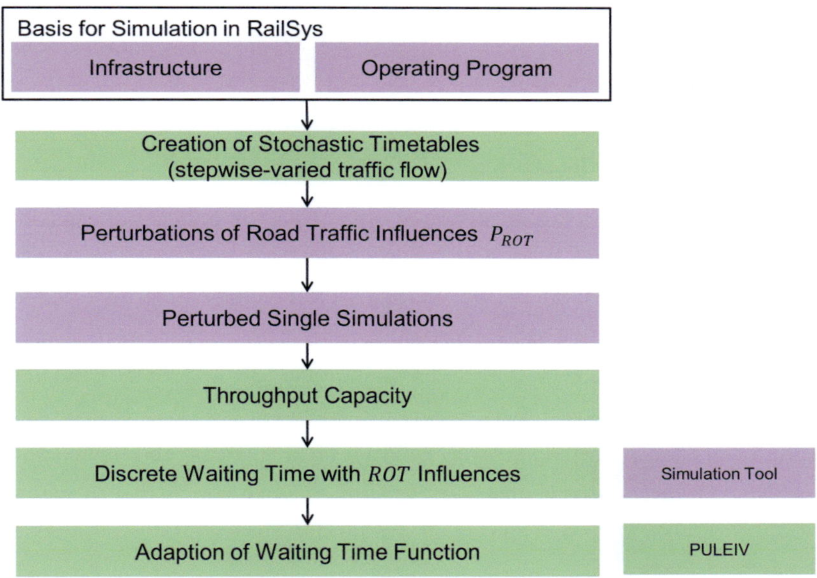

Figure 5-2: Workflow of Capacity Research with the Distribution Approach

Based on the derived running time extension caused by road traffic, and integrated with the existing method for capacity research for railway systems, the capacity research can be carried out as shown in Figure 5-2. A serious of stochastic timetables with stepwise-varied traffic flows for the investigated infrastructure are created based on a given operating program with a selected timetable frame (usually during the high traffic flow time period) with the help of the software PULEIV. A series of perturbed stochastic timetables with the derived perturbation parameters P_{ROT} of the running time extension are generated based on each created stochastic timetable with stepwise-varied traffic loads.

Simulations are carried out with these generated perturbed stochastic timetables with the help of the simulation tool RailSys. The simulation results are loaded to determine the throughput capacity and fit the discrete waiting time curve (data points) with the developed waiting time function (see Chapter 6). The recommended area of traffic flow with consideration of road traffic influences can then be derived.

The result of the investigated example for a large-scale rail-bound network with level crossings influenced by road traffic is shown in Figure 5-3. (The description of the investigated network and the operating program are presented in Subchapter 4.2.1, Appendix Figure 1 and Appendix Table 3). The blue curve is fitted by data points with the developed waiting time function (6-11) with road traffic influences, and the green curve is fitted by the data points with the existing waiting time function (6-5) without road traffic influences (see Chapter 6).

For the situation considering the influences of road traffic, the throughput capacity is 103.1 trains/h, which is much lower than the throughput capacity (146.3 trains/h) for the case without road traffic influences. Similarly, the recommended area of traffic flow is between 63.6 trains/h and 92.2 trains/h for road traffic influences (highlighted in blue). It is lower than the recommended area of traffic flow without the road traffic influences (between 93.9 and 139.8 trains/h, highlighted in green). The differences between the two fitted curves significantly illustrate the influences road traffic has on urban rail-bound transport under various traffic flows.

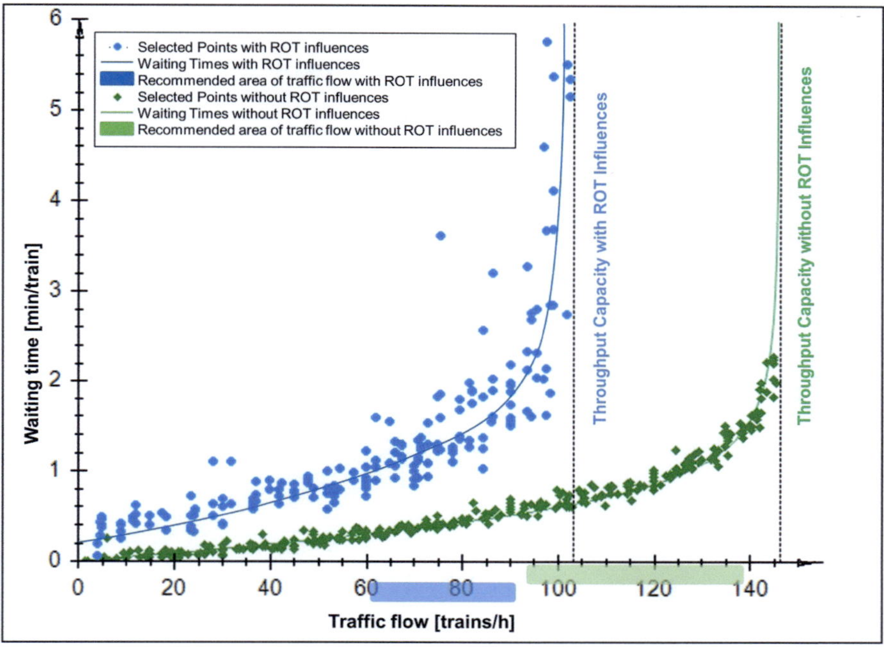

Figure 5-3: Results of the Distribution Approach Comparison to that without Road Traffic Influences for an Investigated Example (Source: modified based on [Martin & Di Liu 2016])

The stochastic timetables with stepwise-varied traffic flows can be simulated with the perturbations of the running time extension based on the perturbation parameters P_{ROT} caused by road traffic. Therefore, the discrete waiting times of urban rail-bound transport resulted from the simulations of urban rail-bound transport include the extra waiting time caused by the road traffic influences. The workflow and the methodology for determining the throughput capacity have been already been discussed in [Chu 2014], and can also be applied to the investigations for urban rail-bound transport in mixed traffic zones considering the influences of road traffic.

However, the existing waiting time function is not appropriate for use with urban rail-bound transport with consideration of road traffic influences. The adapted waiting time function considering the external influences caused by road traffic is developed in Chapter 6. The validation of the results and the comparison between the modeling approach and the distribution approach are discussed in Chapter 7.

6 Derivation of an Adapted Waiting Time Function

6.1 Overview

With consideration of the road traffic influences on urban rail-bound transport, the existing waiting time function for capacity research of urban rail-bound transport is not entirely appropriate. Therefore, an improved model of the waiting time function for capacity research of urban rail-bound transport with consideration of road traffic influences will be derived. The new waiting time function is adapted on the basis of the existing one, but it is much more adaptive to the situation of urban mixed traffic.

If there is only one train in an investigated area of a pure railway system, there are no operational hindrances between trains. The waiting time of this train is equal to zero. However, theoretically for urban rail-bound transport in mixed traffic zones, even though there is only one train (urban rail-bound transport) in an investigated area, it can also be hindered by road traffic in mixed traffic zones. Therefore, extra waiting time for urban rail-bound transport caused by road traffic influences may occur at mixed traffic zones.

An additional variable that is used to reflect the road traffic influences will be studied on the developed waiting time function. The value of parameters in the waiting time function can be approximated by the fitted curve of the discrete waiting time (data points) with road traffic influences from the results of simulations. It is necessary to assess and verify the fit of the new model function of the adapted waiting time function with the discrete waiting time variable.

6.2 Existing Waiting Time Function

This subchapter will describe the existing waiting time function, which has been developed in the research process of capacity research. Further development on the waiting time function will be based on the development process of it in the past and the basic concept of the waiting time function. In this research project, the waiting time function is adapted to the operational situations of investigated areas with consideration of external influences that are caused by road traffic on urban rail-bound transport.

6.2.1   Basic Concept

The waiting time function is an important intermediate result for the two main methods of capacity research, which are the simulation method and the analytic method (see Subchapter 2.3). [Ludwig 1990] and [Hertel 1992] proposed the theory for capacity research of double-track railway lines, which can derive the recommended area of traffic flow using the waiting time function with two parameters. Therefore, the waiting time function plays a very important role in capacity research. [Schmidt 2009] used the simulation method for capacity research to determine the waiting time function with a logarithmic approximation approach. In addition to this, an algorithm to determine the maxim capacity of the investigated area was developed.

Further development of the waiting time function (without consideration of road traffic influences) within the simulation method was carried out by [Martin & Chu 2012] and [Chu 2014]. The effects of the transient phase in simulation were considered. The waiting time function by Chu was completed to determine the throughput capacity and the waiting time function with its 3 parameters. In this research project, a method for the determination of the throughput capacity derived by [Chu 2014] is also used for capacity research of urban rail-bound transport with consideration of road traffic influences.

According to the research by [Chu 2014], the determination of the average of the standard deviation of the difference between the entry traffic flow and the exit traffic flow is related to the number of discrete waiting time data points, which in turn affects the accuracy of the derivation of the throughput capacity. The more the simulated timetables with stepwise-varied traffic flows are available for evaluation. the more exact is the average of standard deviation. Considering the cost for the creation and simulation of these timetables, it is necessary to define the minimum required number of timetables with stepwise-varied traffic flows for a defined accuracy. It was based on [Anderson et al. 2011] to determine the minimum required number of timetables (≈ 80) with stepwise-varied traffic flows through the method Z-test with 5% error probability.

The waiting time function describes the performance of the investigated system and the relationship between the waiting time (quality of operation) and the traffic flow (capacity). In Subchapters 6.2.2 the basic concept for the existing waiting time function is described.

6.2.2 Waiting Time Function by Ludwig and Hertel

As mentioned previously, [Ludwig 1990] and [Hertel 1992] proposed the basic model function of the waiting time function based on the queuing theory. This basic waiting time function with two parameters considered the stationary phase is as follows:

$$ET_w = a \cdot \frac{\eta}{(1 - \eta)^b} \qquad (6\text{-}1)$$

With:

ET_w (statistical) Expectancy value of waiting time function.

a, b Parameters of the waiting time function.

η Utility factor of capacity (traffic flow/maximum capacity).

The mathematical principle is the queuing theory, which studies the behaviors of waiting lines. The railway infrastructure was modeled as a queuing system, in which the trains were modeled as requests with $M/M/1$ queue [Ludwig 1990]. This refers to a queuing system where trains arrive according to a Poisson arrival distribution with rate. It is called the arrival rate, which specifies the average arrival frequency (rate) or the intensity per time interval. The service times are also assumed to be independent and exponentially distributed with parameter μ. It is called the service rate and specifies the average service rate of a single server.

$$\rho = \frac{\lambda}{\mu} \qquad (6\text{-}2)$$

$$ET_w = \frac{\rho}{\mu(1 - \rho)} \qquad (6\text{-}3)$$

With:

ET_w Average waiting time of a single request in the queue.

$\rho(\eta)$ Utility factor.

λ Average arrival frequency (rate).

μ Average service rate of a single server.

ρ can be regarded as the parameter η in the waiting time function (6-1), which is the ratio of consumed capacity, which is calculated as the ratio of actual traffic flow divided by the maximum capacity in the railway system.

For the waiting time function with two parameters derived by [Ludwig 1990], in [Chu 2014], it analyzed and confirmed that the coefficient of determination (R^2) of the waiting time function (6-1) is suitable for simple examples (partial track sections). This indicates that the waiting time function fits very well to the studied cases. Capacity research of large railway nodes or complex infrastructure sections of a railway network is often of much greater significance than that of one partial track section. If you use the analytical method for large railway nodes or complex infrastructure sections requires large computations may be required. [Schmidt 2009] developed a method to determine the maximum capacity and the waiting time function with the simulation method for capacity research. However, for the waiting time function, there can be systematic deviations between the discrete waiting times and the fitted curve (6-1).

Accordingly, [Chu 2014] studied and developed a new waiting time function considering the transient phase in operation This function has a higher level of adaption for railway system with different investigated railway infrastructures and various operating programs . Therefore, the derived recommended area of traffic flow can be much more reliable using Chu's method.

6.2.3   Waiting Time Function by Chu

This subchapter introduces the waiting time function with consideration of the transient phase with three parameters developed by [Chu 2014]. As mentioned above, based on the mathematical principle of queuing theory, the waiting time function by Ludwig and Hertel can be suitably used in the stationary phase through modeling the railway system as a queuing system.

Furthermore, [Chu 2014] proposed that the effects of the transient phase have to be considered in the simulation process and evaluation through capacity research. Many different model functions of the waiting time function were checked, such as the model function of waiting time function with the help of a Taylor expansion as second-order and third-order polynomial functions and exponential function. Through the comparison of the coefficient of determination of different model functions of waiting time function, and further comparison of the adjusted coefficient of determination ($\bar{R}^2$) of different forms of waiting time functions with the different degrees of freedom (second-order polynomial function with three degrees of freedom and third-order polynomial function with four degrees of freedom), it was found that curve progressions of these waiting time functions are not a good fit to the data points (discrete waiting time from simulation). Therefore, [Chu 2014] proposed a developed waiting time function with a higher degree of freedom, which is based on the basic waiting time function (6-1) and that can be used for the stationary phase but also has deviations to the discrete waiting time for large railway stations and complex infrastructure sections of a railway system.

Two functions of the waiting time function with a new parameter c_1 were described to increase the flexibility, which were based on the basic waiting time function (6-1). The model function of the waiting time function (6-4) was developed by [Martin & Chu 2012] and the model function of the waiting time function (6-5) was further developed in [Chu 2014] with consideration of the transient phase in the simulation process for railway systems with a higher (adjusted) coefficient of determination than the basic waiting time function:

$$ET_w = a_1 \cdot \frac{\eta^{c_1}}{(1 - \eta)^{b_1}} \qquad (\text{6-4})$$

With:

ET_w (statistical) Expectancy value of the waiting time function.

a_1, b_1, c_1 Parameters of the waiting time function.

η Utility factor of capacity (traffic flow/throughput capacity).

Through the comparison of the (adjusted) coefficient of determination of the new model function of the waiting time function (6-4) and the basic waiting time function (6-1) with the nonlinear robust least squares of regression method "bisquare" weights to fit (see Subchapter 6.3.2), it was found that the waiting time function with an extra parameter can increase the accuracy of the fitted curve. The model function of the waiting time function (6-4) is better than the basic one (6-1).

Additionally, a further development in [Chu 2014] is with an extra multiplicative term in the basic waiting time function (6-1), and also a new parameter c_2 was used to improve the degree of freedom and adjust the fitted curve of the waiting time function. This model function of waiting time function (6-5) considered the transient phase, and was also based on the queuing theory:

$$ET_w = a_2 \cdot \frac{\eta}{(1-\eta)^{b_2}} \cdot (1 - \eta^{c_2})$$

(6-5)

With:

ET_w	(statistical) Expectancy value of the waiting time function.
a_2, b_2, c_2	Parameters of the waiting time function.
η	Utility factor of capacity (traffic flow/throughput capacity).

The new multiplied term $(1 - \eta^{c_2})$ can be specified as an approximation of the transient phase in the queuing system [Chu 2014]. In the stationary phase, the possible number of trains (requests) in the queuing system n_{req} is considered to be infinite, which is the basis for the basic waiting time function (6-1). However, when the possible number of trains in the queuing system is considered to be finite to N_{req}, the average waiting time of the trains in the queuing system can be considered as an approximation of the transient phase.

$$EL_v = \sum_{n_{req}=0}^{\infty} n_{req} \cdot (1-\rho) \cdot \rho^{n_{req}} \begin{cases} = \dfrac{\rho}{(1-\rho)} & \left(n_{req} \to \infty\right) \\ \approx \dfrac{\rho \cdot (1-\rho^{N_{req}})}{(1-\rho)} & \left(n_{req} \to N_{req}\right) \end{cases}$$

(6-6)

With:

EL_v	Average number of requests in the queuing system.
n_{req}	Possible number of requests in the queuing system.
$\rho(\eta)$	Utility factor.

The multiplied term $(1 - \rho^{N_{req}})$ is equivalent to the relationship of the average number of requests between the transient phase and the stationary phase (see Figure 6-1). $(1 - \rho^{N_{req}})$ was transferred as a multiplied term in the model function of the waiting time function (6-5) with the parameter c_2 instead of N_{req}, which could represent the effects of the transient phase [Chu 2014].

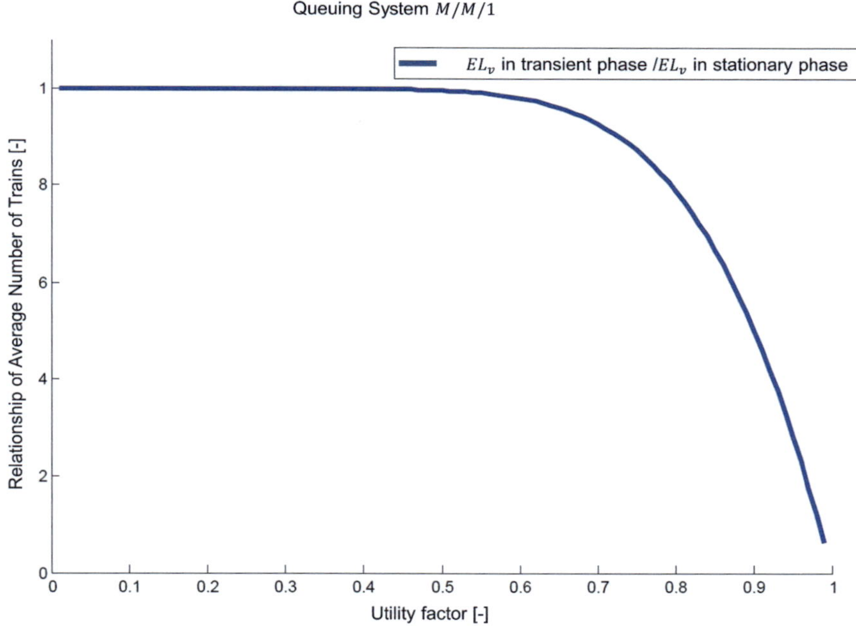

Figure 6-1: Relationship of the Average Number of Trains (Requests) between the Transient Phase and the Stationary Phase (Source: [Chu 2014])

The fit option for adjusting the waiting time function (6-5) with a multiplied term $(1 - \eta^{c_2})$ also uses nonlinear robust least squares with the regression method "bisquare" weights. A comparison of the (adjusted) coefficient of determination of the three waiting time functions (6-1), (6-4), and (6-5) shows that the waiting time functions (6-4) and (6-5) with three parameters of one more degree of freedom can better increase the accuracy of fitted curve with lower deviations from the data points (discrete waiting time) than the basic waiting time function (6-1). Additionally, the waiting time function (6-5) has a better coefficient of determination than the waiting time function (6-4). However, in this research project, the urban rail-bound transport with road traffic influences has to be considered, which means the extra waiting times are generated. The model function of the waiting time function has to be adapted.

6.3 Adapted Waiting Time Function for Urban Mixed Traffic

When road traffic influences on urban rail-bound transport exist in the investigated area, it means that the waiting time of urban rail-bound transport is not only increased by the occurring operational hindrances between trains, but also the influences from other perturbations. Therefore, for urban rail-bound transport with consideration of road traffic, the adapted waiting time function can show the statistical waiting time of the traffic flow with less than and equal to one train should shift up slightly. Moreover, it is necessary to consider the value of the shift for an investigated infrastructure with a given operating program.

Through the acquisition of data points (the discrete waiting times) from the simulation results through simulating the stochastic timetables with stepwise-varied traffic flows of urban rail-bound transport with road traffic influences in mixed traffic zone, the fitted curve of data points with the adapted waiting time function has to be adjusted by suitable fit options. The fit options of nonlinear robust least squares with regression method "bisquare" weights are also used in this research project (see Subchapter 6.3.2).

Even though there are external influences, the adapted waiting time function also needs to specify the waiting time due to the occurring operational hindrances between two trains. The basic model function of the waiting time function (6-1) has to be kept. In consideration of the transient phase as well, the model function of the waiting time function (6-5) developed by [Chu 2014] is used as the basic form for further development. There are several proposed model functions of the waiting time function that will be discussed in the following subchapter.

6.3.1 Adapted Waiting Time Function

In urban mixed traffic patterns, the fitted curve of data points with the existing waiting time function exhibits a deviation from the trend of the discrete waiting times. In this subchapter, new model functions of the adapted waiting time function are discussed, which consider the road traffic influences in investigated areas of urban mixed traffic zones.

The curve of data points with the existing model function for the waiting time function is a statistical result with the starting point of zero (railway system without road traffic influences). When the traffic flow of urban rail-bound transport is in the range (0, 1], it has the high probability of having only one train in the investigated area during the investigated time period, hence there is almost no waiting time due to the occurring operational hindrance between different trains.

For the curve of data points with the adapted waiting time function with road traffic influences, waiting time may occur due to the operational hindrances from the influences of road traffic. Therefore, the starting point of the new waiting time function for urban rail-bound transport shows that the waiting time is more than zero with the influences of road traffic. Therefore, an additional parameter d is considered and studied to be added to the waiting time function (6-5), which represents the starting waiting time caused by road traffic influences in the waiting time function.

In this research, several model functions of the waiting time function with an additional parameter d are studied. In the model function of the waiting time function (6-7), parameter d_3 is a constant added into the existing waiting time function (6-5) and is used to reflect the variation due to the road traffic influences. It is necessary to check whether the (adjusted) coefficient of determination of the adapted waiting time function with parameter d_3 is higher than the corresponding coefficient of the existing waiting time function (6-5).

$$ET_w = a_3 \cdot \frac{\eta \cdot (1 - \eta^{c_3})}{(1 - \eta)^{b_3}} + d_3 \qquad\qquad (\,6\text{-}7\,)$$

With:

ET_w (statistical) Expectancy value of the waiting time function.

a_3, b_3, c_3, d_3 Parameters of the waiting time function.

η Utility factor of capacity (traffic flow/ throughput capacity).

The (adjusted) coefficient of determination of the existing waiting time function (6-5) is compared to that of the adapted waiting time function (6-7) with a constant parameter d_3 , which results from capacity research in the investigated area described in Subchapter 4.2.1 and shown in Table 6-1.

Waiting Time Function		Coefficient of Determination	Adjusted Coefficient of Determination
$ET_w = \dfrac{a_2 \cdot \eta \cdot (1 - \eta^{c_2})}{(1 - \eta)^{b_2}}$	(6-5)	0.9848	0.9847
$ET_w = \dfrac{a_3 \cdot \eta \cdot (1 - \eta^{c_3})}{(1 - \eta)^{b_3}} + d_3$	(6-7)	0.9877	0.9865

Table 6-1: Comparison of the Coefficient of Determination of the Adapted Waiting Time Function (6-7) and the Existing Waiting Time Function (6-5)

The coefficient of determination (R^2) in statistics is a statistical measure to check the regression fit data points. It indicates the goodness of fit of a model (a curve), which is defined by the difference of 1 and the proportion of the total sum of squares $SS(Total)$ by the residual sum of squares $SS(Res)$ with function (6-8) ([Everitt 2002] & [Draper & Smith 1998]). The coefficient of determination (R^2) is in the range of [0, 1]. The better the regression fits the data, the closer the value of coefficient of determination (R^2) is to 1.

$$R^2 = 1 - \frac{SS(Res)}{SS(Total)}$$

(6-8)

With:

R^2 The coefficient of determination.

$SS(Res)$ The residual sum of squares.

$SS(Total)$:   The total sum of squares.

For the waiting time function (6-7), it has a higher degrees of freedom (a_3, b_3, c_3 and d_3) than the existing waiting time function (6-5) (a_2, b_2 and c_2). In order to determine a suitable model function of the waiting time function with a higher explanatory power, the adjusted coefficient of determination is used to compare the different model functions of the waiting time function. The adjusted coefficient of determination ($\bar{R}^2$) is a rescaling of the coefficient of determination by degrees of freedom, which is defined through the proportion of mean of squares $MS(Res)$ and $MS(Total)$ instead of sum of squares with function (6-9) [Draper & Smith 1998].

$$\bar{R}^2 = 1 - \frac{MS(Res)}{MS(Total)} = 1 - \frac{(1 - R^2) \cdot (n_s - 1)}{(n_s - deg. - 1)} \qquad (\,6\text{-}9\,)$$

With:

$\bar{R}^2$ The adjusted coefficient of determination.

n_s The number of sample (data points).

$deg.$ The degrees of freedom of the waiting time function.

$MS(Res)$ The residual mean of squares.

$MS(Total)$ The total mean of squares.

The parameters (a_3, b_3, c_3 and d_3) in the adapted waiting time function (6-7) have to be determined. In order to adjust these parameters of the adjusted waiting time function, the fit option of nonlinear robust least squares with the regression method "bisquare" weights is used for fitted curve. The result of the fitted curve is shown in Figure 6-2. It is clear that the starting point of the waiting time function (6-7) can express the road traffic influences better than the existing waiting time functions (6-1) and (6-5). However, with the waiting time function (6-7), only a new parameter d_3 is added on the existing model function of waiting time function, the deviations between the data points (discrete waiting time) and the fitted curve is also obvious. It cannot well fit the data points with a constant. Especially at higher traffic flows, the fitted curve greatly deviates from the data points with the waiting time function (6-7).

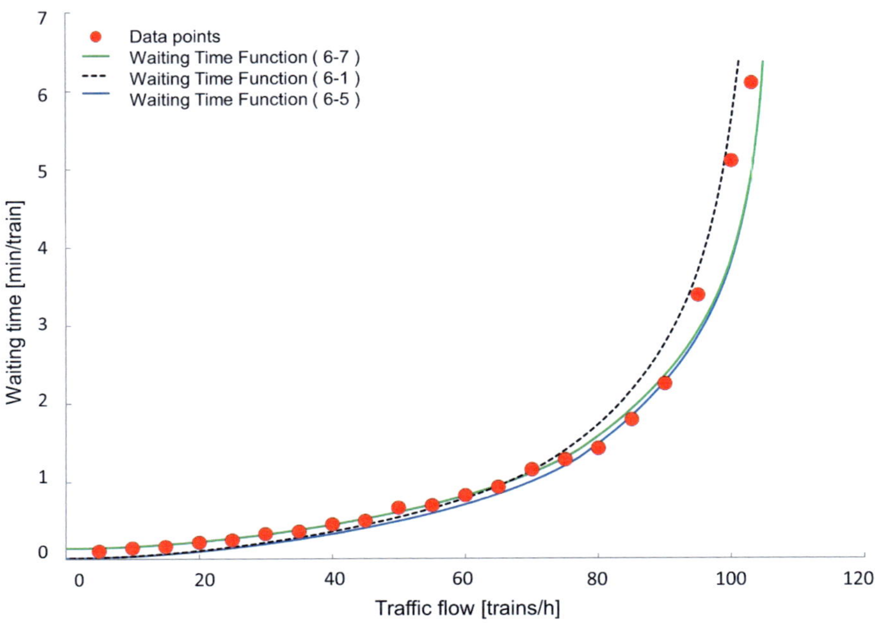

Figure 6-2: Fitting Curve with the Waiting Time Function (6-7) with a Constant Parameter d_3

Further study on the adapted waiting time function is considered to analyze the difference of the fitted curve with the adapted waiting time function (6-7) and the data points (discrete waiting times) from the simulation. Besides, the variation of data points along shows that the waiting time is rapidly increasing with the increase of traffic flow of urban rail-bound transport. Therefore, a new model function for the waiting time function (6-10) with an additional term $d_4/(1-\eta)$ is studied to be added into the waiting time function (6-5), which includes the fourth parameter d_4. The value of the new term $d_4/(1-\eta)$ is proportional to the value of the utility factor of capacity η that is in proportion to the traffic flow of urban rail-bound transport.

$$ET_w = a_4 \cdot \frac{\eta \cdot (1 - \eta^{c_4})}{(1 - \eta)^{b_4}} + \frac{d_4}{(1 - \eta)} \qquad (6\text{-}10)$$

With:

ET_w $\qquad$ (statistical) Expectancy value of the waiting time function.

$a_4, b_4, c_4\ d_4$ $\quad$ Parameters of the waiting time function.

η Utility factor of capacity (traffic flow/throughput capacity).

Similarly, the fit option of the nonlinear robust least squares with the regression method "bisquare" weights is used for adjusting the approximation of the waiting time function (6-10). Figure 6-3 shows the result of fitted curve of data points with the waiting time function (6-10). Through adding a new term $d_4/(1-\eta)$, the fitted curve with waiting time function (6-10) can reflect the data points at the starting point etter, because the new added term is not a constant parameter, but related to the change of the traffic flow. Through the result shown in Figure 6-3, the waiting time function (6-10) can fit the data points better than the waiting time functions (6-5) and (6-7). The deviation between the data points and the fitted curve is only at the middle to high traffic flow.

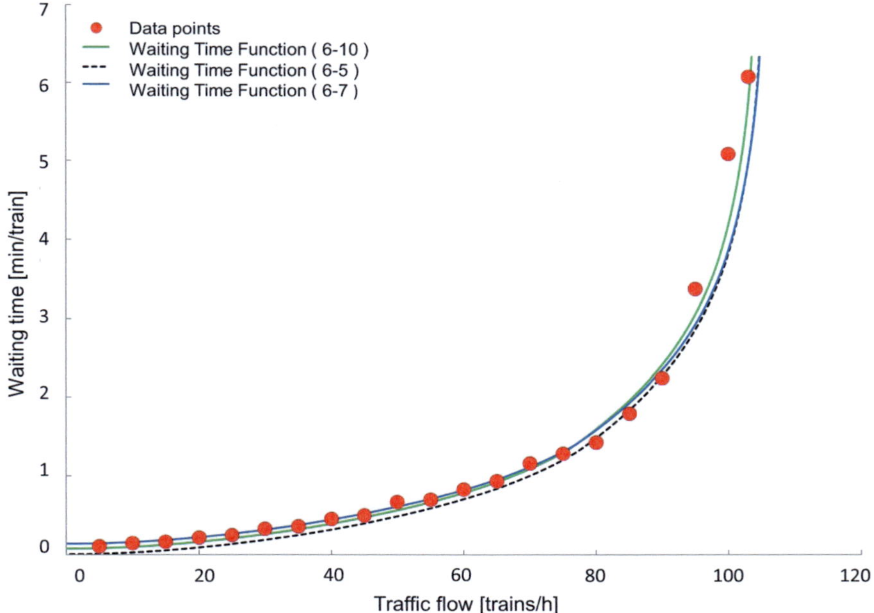

Figure 6-3: Fitting Curve with the Waiting Time Function (6-10) with an additional Term $d_4/(1-\eta)$

The (adjusted) coefficient of determination of the waiting time function (6-10) is shown in Table 6-2. It can be identified that the adjusted waiting time function (6-10) with an additional term of the proportion of parameter d_4 and the difference between 1 and utility factor η is more appropriate than the waiting time function (6-7) with a constant parameter of constant d_3 (see Table 6-1) and clearly better than the existing waiting time function (6-5) with three parameters. Accordingly, the derived recommended area of traffic flow has a higher explanatory power.

Waiting Time Function		Coefficient of Determination	Adjusted Coefficient of Determination
$ET_w = a_2 \cdot \dfrac{\eta}{(1-\eta)^{b_2}} \cdot (1 - \eta^{c_2})$	(6-5)	0.9848	0.9847
$ET_w = a_4 \cdot \dfrac{\eta \cdot (1 - \eta^{c_4})}{(1-\eta)^{b_4}} + \dfrac{d_4}{(1-\eta)}$	(6-10)	0.9901	0.9898

Table 6-2: Comparison of Coefficient of Determination of the Waiting Time Function (6-10) and the Existing Waiting Time Function (6-5)

Through further analysis, the fitted curve of the data points with the waiting time function (6-10) with an additional term $d_4/(1 - \eta)$ for middle to high traffic flow approximates the data points (discrete waiting time) relatively fast. As the traffic flow increases, it is necessary to reduce the increase of the waiting time by introducing an additional term. Therefore, the further adjusted model function of **the** waiting time function (6-11) with an additional term $d_5/(1 - \eta)^{2/3}$ is derived.

$$ET_w = a_5 \cdot \frac{\eta \cdot (1 - \eta^{c_5})}{(1-\eta)^{b_5}} + \frac{d_5}{(1-\eta)^{2/3}} \qquad (6\text{-}11)$$

With:

ET_w	(statistical) Expectancy value of the waiting time function.
$a_5, b_5, c_5\ d_5$	Parameters of the waiting time function.
η	Utility factor of capacity (traffic flow/throughput capacity).

It also applies the method of the nonlinear robust least squares with regression method "bisquare" weights and the Trust-Region-Algorithm for the determination of the parameters for the adapted waiting time function (6-11) with data points (discrete waiting times).

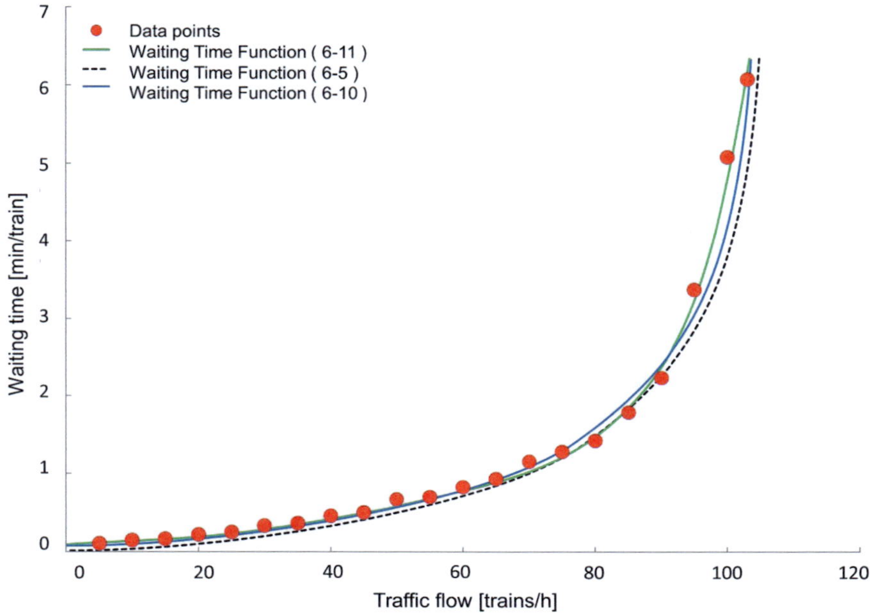

Figure 6-4: Fitted Curve with the Waiting Time Function (6-11) with an additional Term $d_5/(1-\eta)^{2/3}$

Figure 6-4 indicates the fitted curve of the data points with the new adapted waiting time functions (6-10) and (6-11). The fitted curve with waiting time function (6-11) has almost no deviations from the data points. Compared with the results from the waiting time function (6-10), the new added term $d_5/(1-\eta)^{2/3}$ adjusts the changing state from the starting point with the increase of traffic flow. Even with middle to high traffic flow, the fitted curve of data points with the adapted waiting time function (6-11) also fits the data points well.

Waiting Time Function		Coefficient of Determination	Adjusted Coefficient of Determination
$ET_w = \dfrac{a_4 \cdot \eta \cdot (1 - \eta^{c_4})}{(1 - \eta)^{b_4}} + \dfrac{d_4}{(1 - \eta)}$	(6-10)	0.9901	0.9898
$ET_w = \dfrac{a_5 \cdot \eta \cdot (1 - \eta^{c_5})}{(1 - \eta)^{b_5}} + \dfrac{d_5}{(1 - \eta)^{2/3}}$	(6-11)	0.9928	0.9927

Table 6-3: Comparison of the Coefficient of Determination of the Adapted Waiting Time Functions with the Four Parameters (6-10) and (6-11)

As shown in Table 6-1, the waiting time function (6-5) developed by [Chu 2014] with three parameters has an insufficient (adjusted) coefficient of determination with consideration of road traffic influences. The (adjusted) coefficient of determination on (6-7) with an additional parameter d_3, by contrast, is relatively better. The two model functions of adapted waiting time functions shown in Table 6-3 both include an additional term with the fourth parameter d, which can reflect the changes of the waiting time with the increase of traffic flow of urban rail-bound transport better. Especially, the last adapted waiting time function (6-11) has the relatively best coefficient of determination. Therefore, it is much more plausible that the recommended area of traffic flow is derived with the adaptive waiting time of function (6-11).

6.3.2   Fit Options

The fit options for curve fitting play an important role in deriving the coefficients of determination of the different waiting time functions. The fit options that are used in this research project for the curve fitting of data points with a waiting time function are described in this subchapter.

In general, the method of least squares is used for the fitted curve [Moler 2008]. It is a standard approach in robust regression that is a form of regression analysis, which minimizes the sum of squared errors (residuals in statistics) to get the best fit of data points. A residual is the difference between an observed value and the fitted value derived from a model.

$$res_i = y_i - \hat{y}_i \tag{6-12}$$

$$SS(Res) = \sum_{i=1}^{m} res_i^2 = \sum_{i=1}^{m} (y_i - \hat{y}_i)^2 \tag{6-13}$$

With:

res_i:	The residual for the i th data point.
y_i	The i th of m observed value (observations).
$\hat{y}_i$	The i th of m fitted value with the model function,
$SS(Res)$	The residual sum of squares. /The summed square of residuals where m is the number of data points included in the fit.

x is considered as the independent variable, $y(x)$ denotes the function of x that is used to approximate. The function $y(x)$ is modeled by a combination of n basis functions $\phi_j(x_i, \alpha)$ that involve linear parameters $\beta = (\beta_1, \beta_2, ..., \beta_n)$ and nonlinear parameters $\alpha = (\alpha_1, \alpha_2, ..., \alpha_p)$. The function with the fitted value $\hat{y}_i$ is approximately equal to the observed value:

$$\hat{y}_i(x) \approx \beta_1 \phi_1(x, \alpha) + \beta_2 \phi_2(x, \alpha) + \cdots + \beta_n \phi_n(x, \alpha) \tag{6-14}$$

$$x_{i,j} = \phi_j(x_i, \alpha) \tag{6-15}$$

With:

α	The nonlinear parameters of the model function $\hat{y}_i(x)$.
β	The linear parameters of the model function $\hat{y}_i(x)$.
x_i	The variable of the model function $\hat{y}_i(x)$.
ϕ_j	The j th of n basis function of x.

The residuals of the data points that are the difference between the observations and the model are calculated as:

$$res_i = y_i - \hat{y}_i = y_i - \sum_{j=1}^{n} \beta_j \phi_j(x_i, \alpha) \tag{6-16}$$

The parameters have to be determined with the method of the least squares to derive the minimum of the residual sum of squares (the difference between an observed value and the fitted value of model function). It can be expressed though the residuals using the Euclidean distance in (6-17).

$$\|res\|^2 = SS(Res) = \sum_{i=1}^{m} res_i^2$$ (6-17)

The outliers for the method of least squares are the main disadvantage, which have a large influence on the fit. The robust regression method with "bisquare" weights can be used to minimize the influences of the outliers, which minimizes the weighted sum of squares of the outliers. The weight given to each data point depends on how far the point is from the fitted line. Points near the fitted line get full weight. On the contrary, points farther from the line get reduced weight.

The robust fitting with "bisquare" weights is an iterative procedure done by reweighting the least squares algorithm until the fit converges. The adjusted residuals r_{adj} are given with the usual least squares residuals r_i in model function (6-18) [Venables & Ripley 1997]:

$$r_{adj} = \frac{res_i}{\sqrt{1 - hl_i}}$$ (6-18)

With:

r_{adj} The usual least squares residuals.

hl_i: The leverages that adjust the residuals by reducing the weight of high leverage data points.

The adjusted residuals r_{adj} have to be standardized to compute the robust weights.

$$u = \frac{r_{adj}}{Ks_{rob}} = \frac{r_{adj}}{4{,}685 \cdot \frac{MAD}{0{,}6\ 745}}$$ (6-19)

With:

u The standardized adjusted residuals.

K A tuning constant equal to 4.685.

s_{rob} The robust variance given by $MAD/0{,}6745$.

MAD: The median absolute deviation of the residuals.

Therefore, the robust weights $w_{bisqure}$ of each iterative are given by:

$$w_i = \begin{cases} (1 - (u_i)^2)^2 & |u_i| < 1 \\ 0 & |u_i| \geq 1 \end{cases}$$

(6-20)

With:

w_i The robust bisquare weights $w_{bisqure}$ of i th iterative.

u_i The standardized adjusted residuals of i th iterative.

In this research project about the fitted curve of the waiting time function, the main model function is nonlinear. Due to the coefficients (variables) for the waiting time function that cannot be estimated using a simple matrix technique, the nonlinear models are much more difficult to fit than linear models. The iterative approach is required for nonlinear least squares. In the process, it is necessary to adjust the coefficients and adjust whether the fit improves. The direction and magnitude of the adjustment depend on the selected fitting algorithm. The Trust-Region-Algorithm is an algorithm for mathematical optimization and is described in [Sorensen 1982] and [Moré & Sorensen 1983]. It is the default algorithm in the curve fitting toolbox in Matlab[15], which is used if the lower or upper coefficient constraints are specified.

The software Matlab is used in this research project. In the curve fitting toolbox, the fit option of the robust least squares with the regression method "bisquare" weights and Trust-Region-Algorithm are selected for the curve fitting of the waiting time function. As shown in Figure 6-5, the adjusted coefficient of determination of the green fitted curve of the waiting time function (6-11) with the nonlinear robust least squares with regression method "bisquare" weights is 0.9927, which is higher than the blue fitted curve of waiting time function without robust regression, 0.9873 in this example. Other than the outlier data point (traffic flow with about 70 trains/h), the green curve is a better fit for all of the data points. The outlier of the data point is detected through robust regression and receives a lower weight through the "bisquare" weights. It leads to a small rise and large deviation between the blue fitted curve and data points without the robust regression.

---

[15] Matlab 2014a Moler [2008]

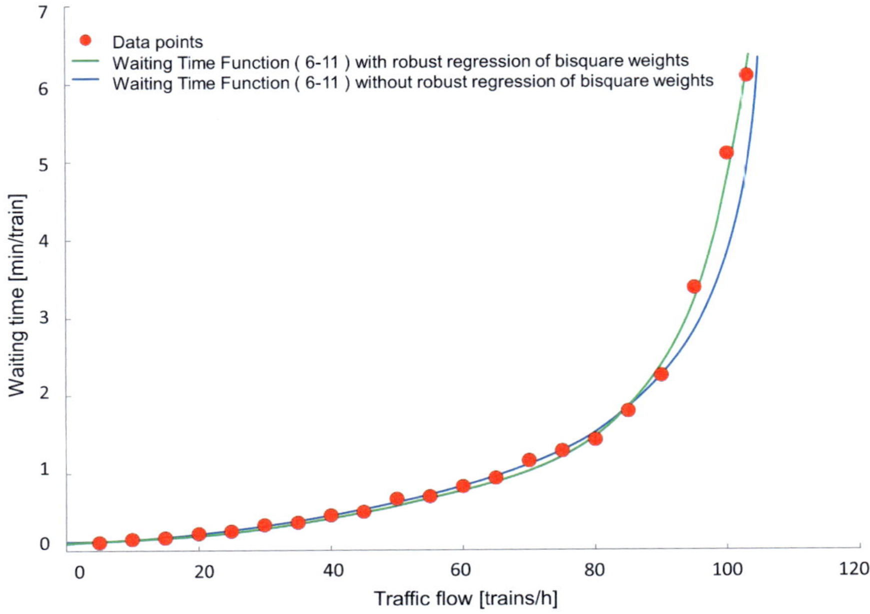

Figure 6-5: Comparison of the Fitted Curve of the Waiting Time Function (6-11) with and without Robust Regression of "Bisquare" Weights

6.4 The Algorithm for the Derivation of the Adapted Waiting Time Function with Road Traffic Influences

In railway systems, the waiting times (without consideration of road traffic influences) result from the hindered trains during operations. With the existing approaches, the waiting time function of railway transportation can be derived with the help of railway simulation tools and software for capacity research, such as RailSys and PULEIV. For urban mixed traffic, due to the additional external influences of road traffic, the efforts of derivation of the improved waiting time function (6-11) are considerably high with simulation methods for capacity research. It is even impossible to derive the waiting time function if the required road traffic information is not fully available. Therefore, an algorithm for deriving the waiting time function is developed in this chapter.

6.4.1 Basic Concept

Due to the difficulty of acquiring road traffic information for the operations in mixed traffic zones, this algorithm is developed to approximate the waiting time function (6-11) of urban rail-bound transport with consideration of road traffic influences. Based on the results of capacity research of the investigated mixed traffic zone without consideration of road traffic influences, the waiting time function (6-5) can be determined with three parameters (a_2, b_2, c_2) using the simulation method without road traffic influences for the investigated mixed traffic zone.

If the road traffic influences are taken into account, the throughput capacity $(\widehat{DS\,LF})$ will generally decrease. The mixed traffic zone with road traffic influences can be regarded as a bottleneck in the investigated area. In addition, the parameter d_5 in the waiting time function (6-11) describes the road traffic influences when there is only one urban rail-bound transport vehicle in the investigated area. In this situation, it is assumed that the three parameters (a_2, b_2, c_2) will not be changed when introducing road traffic influences. With this assumption, the three parameters are determined in a simplified way based on the results of capacity research carried out for pure rail-bound systems without road traffic influences.

Therefore, in order to derive the approximate waiting time function considering road traffic influences in the investigated mixed traffic zone, two important parameters have to be determined:

- $\widehat{DS\,LF}$: Throughput capacity with road traffic influences.
- $d\ (= d_5/(1 - \eta)^{2/3})$: The starting point (waiting time) of the waiting time function (6-11) with road traffic influences.

According to the rules of traffic control signaling systems, the operations of urban rail-bound transport and road traffic at level crossings reflected by the phase rotation of traffic control signals can be described as an event-driven system. Therefore, the waiting times of the influenced urban rail-bound transport can be calculated.

Through the application of the automatic event-driven system, enormous efforts can be reduced, when compared to manual calculation. Moreover, the acquisition of the operational road traffic information can be simplified to a great extent. In Subchapter 6.4.2, the algorithm and the relevant input data will be specified. The various cases in the event-driven system and the workflow to determine the previously-mentioned two parameters for the derivation of the waiting time function will be discussed in Subchapter 6.4.3 and 6.4.4. At last, the further application for a large-scale investigated area will be introduced in Subchapter 6.4.5.

6.4.2   Algorithm Description

According to the conditions of the investigated mixed traffic zone, the movements of road traffic are modeled by the rotation and durations of the traffic light phases in the traffic cycle during the investigated time period. The interfered urban rail-bound transport are semi-randomly generated and inserted into the system according to the traffic flows in the scheduled timetable, which represent the urban rail-bound transport arriving at the mixed traffic zone. At the starting point in time of its/their arrival(s), the interactions between the arriving urban rail-bound transport and the road traffic are modeled in the event-driven system through triggering events. Therefore, the influences of the road traffic on the urban rail-bound transport can be reflected in the modeling of an event-driven system.

According to [Chu 2014], in one determined time slice exactly on the second, there is at least one urban rail-bound transport vehicle created in the stochastic timetables with stepwise-varied traffic flows based on the operating program. The time slice is determined by the traffic flow. For example, if the traffic flow is 6 trains/h in the investigated time period of six hours, the time slice of 10 minutes can be determined by using the following formula:

$$T_{sl} = \frac{T \cdot 60 \cdot 60}{\sum_{t=1}^{T} N(t)} \qquad (6\text{-}21)$$

With:

T_{sl}	The time slice in seconds.
T	The investigated time period in hours:
t	The index of the investigated time period, $t \in [1, \ T]$.
$N(t)$	The traffic flow of the urban rail-bound transport routes in each hour of the investigated time period.

It means there is one urban rail-bound transport vehicle every ten minutes and the arriving time of this vehicle can be created at any (random) point in time in this time slice.

In Figure 6-6, the general process of modeling an event-driven system in the algorithm is illustrated. Firstly, the basic information of the investigated mixed traffic zone, the infrastructure, the given operating program (scheduled timetable) of urban rail-bound transport, as well as the rotation and durations of traffic light phases of traffic control signals are the prerequisites for modeling. The relevant rules have to be set at the beginning as well. Afterwards, during the investigated time period, the urban rail-bound transport are inserted according to the stepwise-varied traffic flow, which may interrupt the pre-defined normal rotation of the traffic light phases of the road traffic at the investigated mixed traffic zone without urban rail-bound transport. Furthermore, when an event is triggered in the event-driven system, one of the various cases will be assigned as a response.

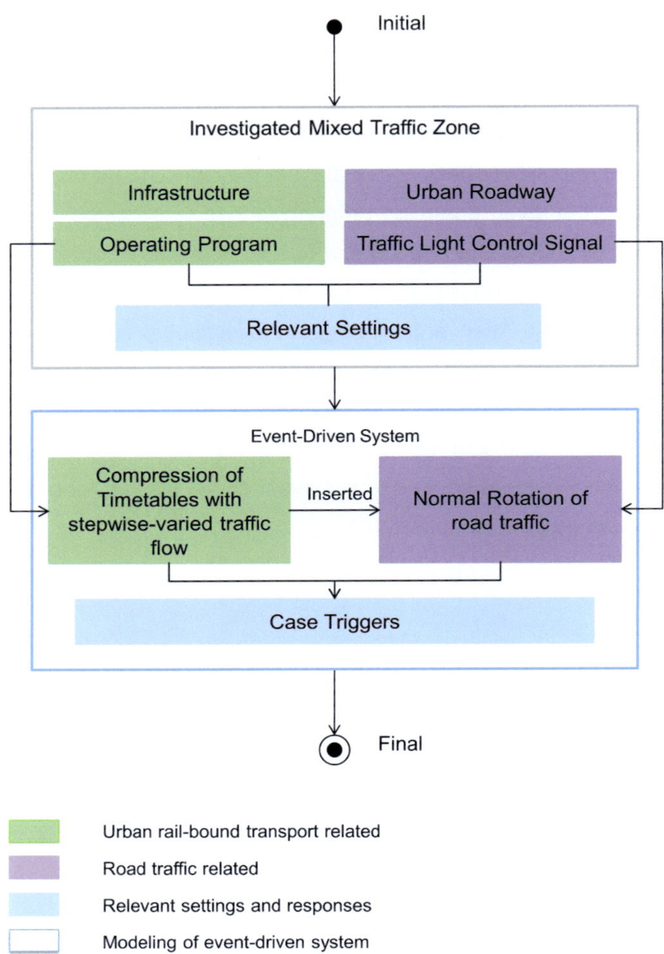

Figure 6-6: Scheme of the Modeling of Event-Driven System based on the Algorithm

Urban rail-bound transport is also influenced by the road traffic, so that the waiting time rises and therefore the blocking time to pass through the mixed traffic zone differs from that in the schedule. The influences of road traffic can be obtained through the hindrance time of urban rail-bound transport by road traffic. The mixed traffic zone of shared roads is the extension of the modeling of mixed traffic zone for level crossings. The mixed traffic zone of shared roads will be discussed later in this sub-chapter.

Input Settings and Relevant Rules

For the investigation of urban mixed traffic zones, the urban rail-bound transport has a relatively higher priority in the operation. However, as described in Chapter 3, there are two prerequisites that have to be fulfilled preferentially.

- The maximum red light phase for road traffic
- The minimum green light phase for road traffic

On the basis of this, the urban rail-bound transport can be prioritized to occupy the mixed traffic zone on other conditions. At mixed traffic zones, the **route** is defined as a section of the infrastructure for urban rail-bound transport or an urban roadway for road traffic with a given direction where the mixed traffic can move from an entry point to an exit point. The duration of a movement of one route is controlled by a traffic control signal in its corresponding phase for road traffic and the blocking time for urban rail-bound transport. It can be subdivided into:

Routes of road traffic

$$R_{ROT} = \{r_{ROT,1}, r_{ROT,2}, \cdots, r_{ROT,p}, \cdots, r_{ROT,m}\} \tag{6-22}$$

Routes of urban rail-bound transport

$$R_{URT} = \{r_{URT,1}, r_{URT,2}, \cdots, r_{URT,q}, \cdots, r_{URT,n}\} \tag{6-23}$$

With:

m	The number of road traffic routes with $|R_{ROT}| = m$.
n	The number of urban rail-bound transport routes with $|R_{URT}| = n$.
p	The index of the road traffic routes, $p \in [1, m]$.
q	The index of the urban rail-bound transport routes, $q \in [1, n]$.

In Figure 6-7 a), examples of the routes in the mixed traffic zones of level crossings and shared roads are described on a microscopic scale for the investigated example. For mixed traffic zones of level crossings (see Figure 6-7 b)), the start point is set as the point of the traffic control signal facility entering the level crossing in the direction of the route, either for routes of urban rail-bound transport or road traffic, and the end point is also considered. For level crossings, (according to the required signaling block system) the end point can be regarded as the point of the next adjacent traffic control signal or can be set at an assumed point such as the end of the block of the level crossing.

For shared roads (see Figure 6-7 c)), a traffic control signal is assumed to be the start point of the shared road in the model. If the adjacent shared road connects with a level crossing, the end point of the route on the level crossing is regarded as the start point of the following routes on the shared road. Furthermore, the end point of the route on the shared road is the start point of the route on the following adjacent level crossing.

For example, the urban rail-bound transport route is the section of infrastructure with direction $(W \rightarrow E)$ from start point s_1 until end point s_4 (assumed point), which is denoted as r_{URT}: $s_1 \rightarrow s_4$. The road traffic route, is a section of urban road with direction $(E \rightarrow S)$ from start point s_5 until end point s_2 (assumed point), which is denoted as r_{ROT}: $s_5 \rightarrow s_2$.

Moreover, on the shared road, the route of mixed traffic is also defined as the section starting from the point at the traffic control signal facility until the end point at the next traffic control signal, which is the start point of another route at the following adjacent level crossing in the same direction. In the example, the road traffic route can be denoted as r_{ROT}: $s_4 \rightarrow s_9$ on the shared road with direction $(E \rightarrow W')$, which starts from the end point s_4 of the route r_{ROT}: $s_7 \rightarrow s_4$ on the adjacent level crossing in front until the end point s_9 on the shared road. Similarly, the route of the urban rail-bound transport with direction $(E \rightarrow W')$ can be denoted as r_{URT}: $s_4 \rightarrow s_9$ with start point s_4 and end point s_9, which is the start point of another route on the following adjacent level crossing.

a)

s_7
s_6
s_8
s_5
s_{10}
s_1
s_4
s_9
s_2
s_3

b)

N
s_6
s_7
s_8
$r_{ROT}: s_5 \rightarrow s_2$
s_5
W
E
$r_{URT}: s_1 \rightarrow s_4$
s_1
s_4
s_3
s_2
S

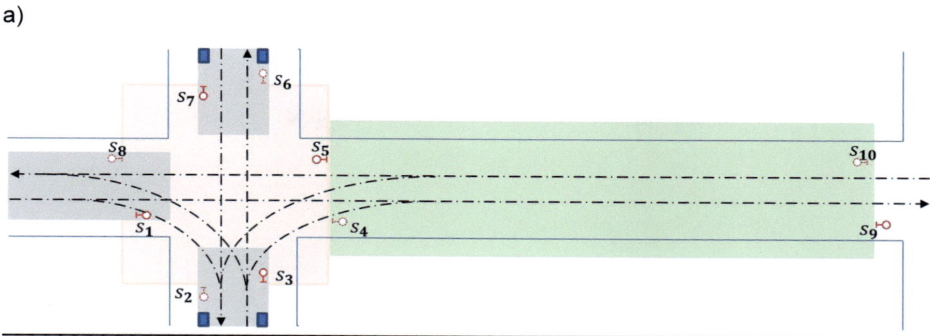

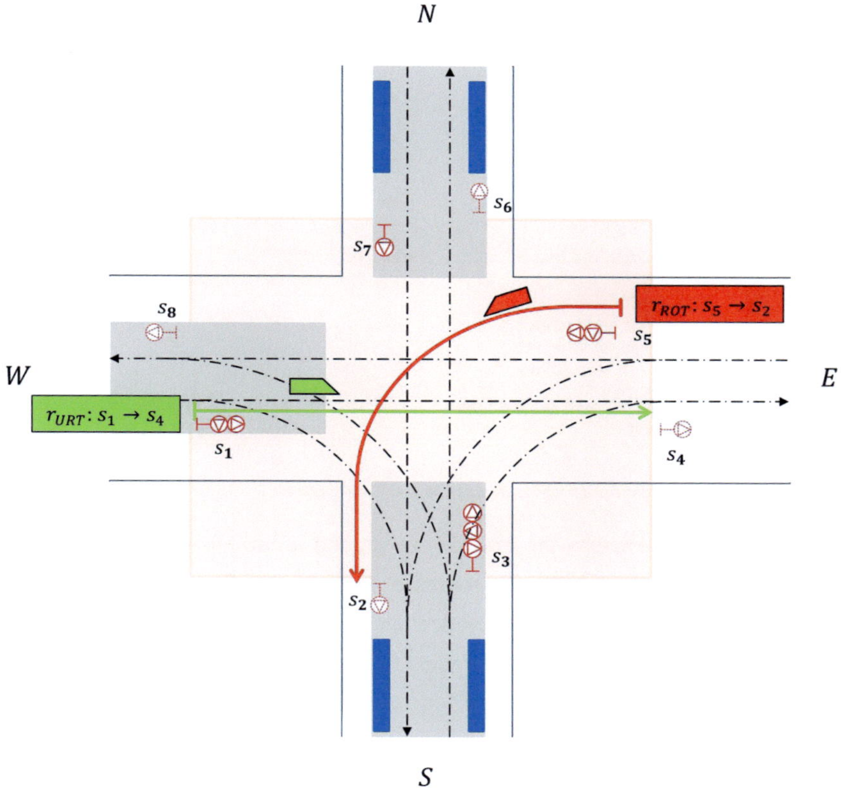

c)

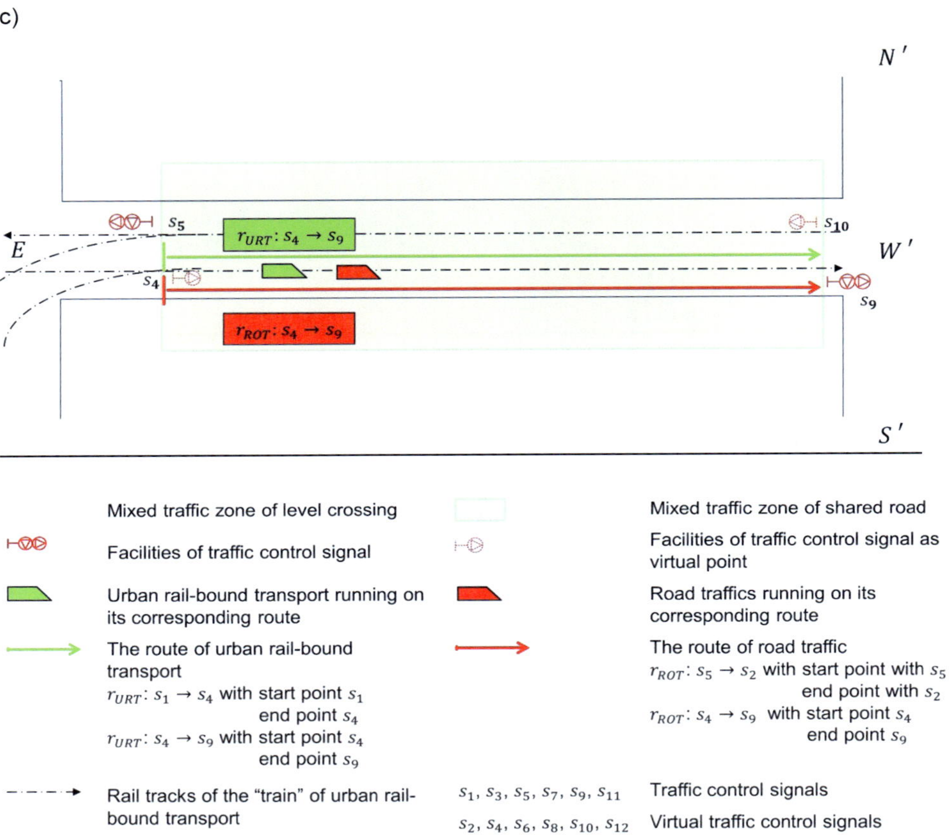

	Mixed traffic zone of level crossing			Mixed traffic zone of shared road
	Facilities of traffic control signal			Facilities of traffic control signal as virtual point
	Urban rail-bound transport running on its corresponding route			Road traffics running on its corresponding route
	The route of urban rail-bound transport $r_{URT}: s_1 \rightarrow s_4$ with start point s_1 end point s_4 $r_{URT}: s_4 \rightarrow s_9$ with start point s_4 end point s_9			The route of road traffic $r_{ROT}: s_5 \rightarrow s_2$ with start point with s_5 end point with s_2 $r_{ROT}: s_4 \rightarrow s_9$ with start point s_4 end point s_9
	Rail tracks of the "train" of urban rail-bound transport	$s_1, s_3, s_5, s_7, s_9, s_{11}$	Traffic control signals	
		$s_2, s_4, s_6, s_8, s_{10}, s_{12}$	Virtual traffic control signals	

Figure 6-7: Microscopic Description of Routes at a Mixed Traffic Zone

In order to explicitly express the interactions among urban mixed traffic (road traffic and urban rail-bound transport), the routes of the mixed traffic and the relationships among them have to be defined as the first step as a reference in a locking table for route related locking for urban mixed traffic.

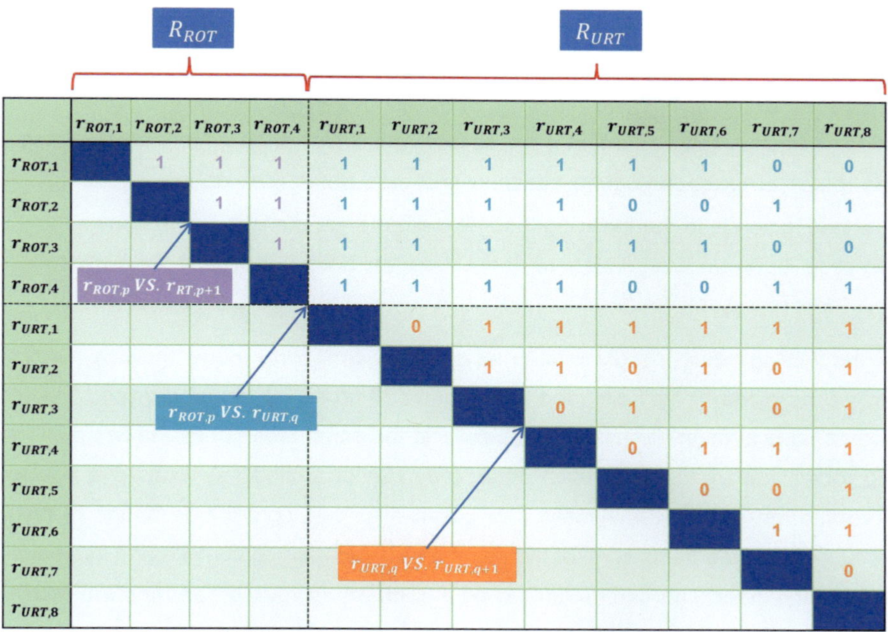

	$r_{ROT,1}$	$r_{ROT,2}$	$r_{ROT,3}$	$r_{ROT,4}$	$r_{URT,1}$	$r_{URT,2}$	$r_{URT,3}$	$r_{URT,4}$	$r_{URT,5}$	$r_{URT,6}$	$r_{URT,7}$	$r_{URT,8}$
$r_{ROT,1}$		1	1	1	1	1	1	1	1	1	0	0
$r_{ROT,2}$			1	1	1	1	1	1	0	0	1	1
$r_{ROT,3}$				1	1	1	1	1	1	1	0	0
$r_{ROT,4}$					1	1	1	1	0	0	1	1
$r_{URT,1}$						0	1	1	1	1	1	1
$r_{URT,2}$							1	1	0	1	0	1
$r_{URT,3}$								0	1	1	0	1
$r_{URT,4}$									0	1	1	1
$r_{URT,5}$										0	0	1
$r_{URT,6}$											1	1
$r_{URT,7}$												0
$r_{URT,8}$												

Table 6-4: The Locking Table of Route Related Locking for Urban Mixed Traffic of Example Mixed Traffic Zone (Source: modified based on [Martin & Di Liu 2016])

The table of route related locking for urban mixed traffic shows the dependency of the routes for the example of a mixed traffic zone with a level crossing as described in Chapter 4.2.1 (see Figure 4-1). The constituent possible number of combinations of two elements from a set of all routes for mixed traffic with number of $m + n$ is shown in Table 6-4.

$$C_2^{m+n} = \frac{(m + n)!}{2!\,(m + n - 2)!}$$

(6-24)

With:

m The number of the road traffic routes with $|R_{ROT}| = m$.

n The number of the urban rail-bound transport routes with $|R_{URT}| = n$.

In this example, there are $C_2^{m+n} = C_2^{4+8} = 66$ combinations to denote the interactions between each two routes of mixed traffic at the investigated mixed traffic zone.

This table is divided into three regions based on the types of routes. The first region indicates the interactions between the routes of the road traffic itself (between $r_{ROT,p}$ and $r_{ROT,p+1}$), which is marked in violet in the top left corner of the table. The middle region with blue content is the second one, which shows the relation between the routes of urban rail-bound transport and road traffic (between $r_{ROT,p}$ and $r_{URT,q}$). The last region is in the lower right corner with orange content, which shows the interaction between the routes of the urban rail-bound transport itself (between $r_{URT,q}$ and $r_{URT,q+1}$).

In the locking table of route related locking, two conflicting routes indicate the two routes (no matter if they are road traffic routes or urban rail-bound transport routes) cannot pass through the mixed traffic zone at the same time. Since the two conflicting routes are exclusive of each other, they can be defined as **exclusive routes combination (ERC)**. Conversely, it is possible for two conflict-free routes to pass through the mixed traffic zone at the same time, which are compatible and independent of each other. This can be defined as a **compatible routes combination (CRC)**. In the investigated example, each pair of the four road traffic routes are exclusive routes combination (ERC) (see Table 6-4). They cannot occupy the mixed traffic zone at the same time.

a)

N
$r_{ROT,1}: s_5 \rightarrow s_2$
$r_{URT,1}: s_1 \rightarrow s_4$
W
E
s_6
s_7
s_8
s_5
s_1
s_4
s_2
s_3
S

b)

N
$r_{ROT,1}: s_5 \rightarrow s_2$
$r_{URT,7}: s_3 \rightarrow s_4$
W
E
s_6
s_7
s_8
s_5
s_1
s_4
s_2
s_3
S

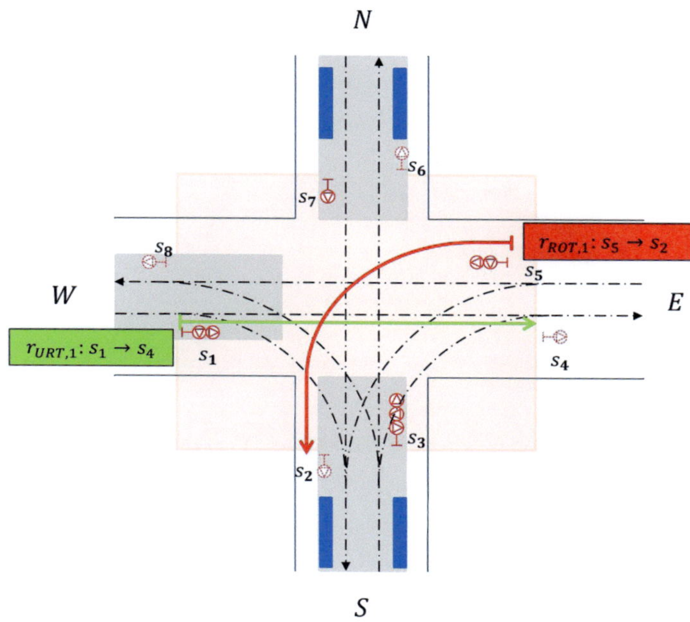

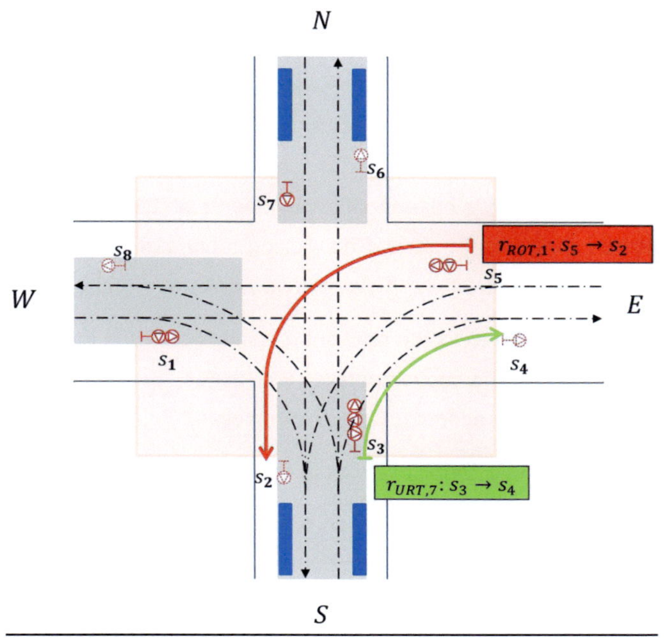

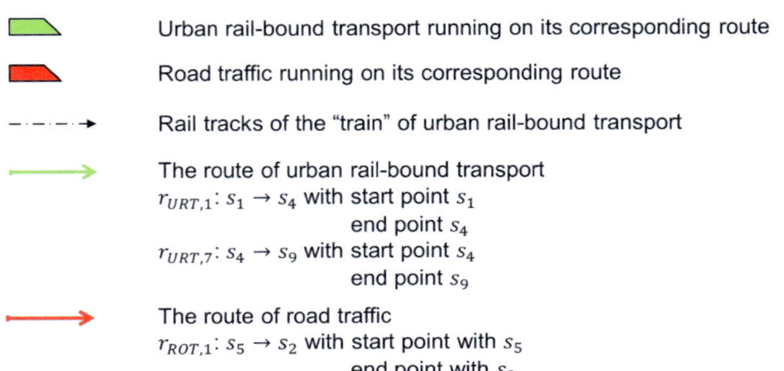

Figure 6-8: An Example of Exclusive Routes Combination and Compatible Routes Combination in the investigated Example

The exclusive routes combinations are marked as "1" in the locking table of route related locking. For instance, the partial routes of the investigated example are shown in Figure 6-8 a). The road traffic route $r_{ROT,1}$ (red arrow line) is in conflict with the urban rail-bound transport route $r_{URT,1}$, (green arrow line). It indicates the combination of these two routes is exclusive. Thus, the exclusive route combination is marked as "1" in the table.

Correspondingly, the compatible routes are marked as "0". The road traffic route $r_{ROT,1}$ and the urban rail-bound transport route $r_{URT,7}$ operate without conflict. This can therefore be identified as a compatible route combination and is marked in the table as "0" (Figure 6-8 b)). As a reference, the locking table of route related locking shows the possibility of simultaneous mixed traffic movements on their corresponding routes at an investigated mixed traffic zone. The locking table can guide the routes of the urban mixed traffic modeled within an event-driven system.

Additionally, it is necessary to fix the relevant rules of routes and traffic control signaling systems that control the movements of the routes, as input in this algorithm. There are mainly three operational rules in this principle.

1) **Sequential rule** for a traffic control signaling system at a level crossing

    The sequence of green light rotation of traffic control signals for road traffic at an investigated level crossing commands the sequence of road traffic routes to proceed in the denoted direction passing through the level crossing in one traffic cycle.

i.   Ordering principle

The green light of traffic control signals for road traffic rotates according to the pre-defined sequence, which has to be kept from one traffic control signal cycle to the next. For example, green light (G) rotation among road traffic routes (R_{ROT}) with a defined sequence[16]:

$$G\left(r_{ROT,1}\right) \to G\left(r_{ROT,2}\right) \to \cdots \to G\left(r_{ROT,p}\right) \to \cdots \to G\left(r_{ROT,m}\right) \to$$
$$G\left(r_{ROT,1}\right) \to G\left(r_{ROT,2}\right) \to \cdots \to G\left(r_{ROT,p}\right) \to \cdots \to G\left(r_{ROT,m}\right) \to \cdots$$

ii.   Disordering principle

The green light phase of traffic control signals for road traffic can rotate to any road traffic routes depending on the requirements. The sequence of the green light phase of traffic control signals for road traffic can be changed in the operation process.

Furthermore, it can be applied for the vehicle-depended traffic control signaling system. The phase of traffic control signals of road traffic can also be adjusted according to the actual traffic flow, which correspondingly changes the occupation time of the road traffic route.

2) **Application rule** for urban rail-bound transport

At urban mixed traffic zone level crossings, the urban rail-bound transport has the right to apply to arrive at the traffic control signal in advance. Before the traffic control signal, the urban rail-bound transport can apply through the device along the rail tracks (see Chapter 3.2.1). The distances between the applying point and the traffic control signal vary between the various urban mixed traffic zones.

In this algorithm, the applying point is assumed to be defined as the start point of the urban rail-bound transport route, at which the urban rail-bound transport

---

[16] The sequence shows here is just one example, the subscript "$1, 2, \cdots, p, \cdots m$" only indicate the element of route set without the meaning of sequence.

can apply for entering and occupying the mixed traffic zone level crossing. Meanwhile, the event-driven system can model the interactions between road traffic and urban rail-bound transport at mixed traffic zones (level crossing) under certain conditions (Chapter 6.4.4).

3) **Limitations of traffic light phase rule** for road traffic

As mentioned, two prerequisites (the minimum green light phase and the maximum red light phase for road traffic at mixed traffic zone level crossings) have to be ensured in the model. Moreover, their values have to be given as input into this algorithm for event-driven system according to the requirements. These values can come from actual data from on-site measurements or from the local urban transportation company.

For various scenarios of different mixed traffic zones and various scenarios of one single mixed traffic zone, the road traffic movements based on the locking table of route related locking can be controlled by traffic control signals with various rotations. One or more of the different compatible routes combination (CRC) of road traffic routes can be defined under one green light phase of the normal rotation in one traffic cycle at the level crossing. There are various possible rotations for one traffic cycle. Therefore, it is important and necessary to determine the rotation of the traffic light phases based on the pre-defined rules and the locking table of route related locking.

In the investigated example, the four road traffic routes are exclusive of each other. If a road traffic route is occupied, the other three road traffic routes have to be in their red light phases. The normal rotation is set with a defined and fixed sequence under the ordering principle:

$$G\left(r_{ROT,1}\right) \to G\left(r_{ROT,2}\right) \to G\left(r_{RTO,3}\right) \to G\left(r_{ROT,4}\right) \to G\left(r_{ROT,1}\right) \to G\left(r_{ROT,2}\right) \to \cdots$$

In addition, the different phases of a road traffic route can be controlled by the corresponding traffic control signal, which can represent the **effective occupation time of the road traffic route**. The effective occupation time ($o_{ROT,p}$) is regarded as the time duration of one road traffic route ($r_{ROT,p}$) starting from its entry until allowing another exclusive mixed traffic route (road traffic or urban rail-bound transport route) to enter the mixed traffic zone. Therefore, the different phases of the road traffic route have to be determined as the input of the model in this algorithm.

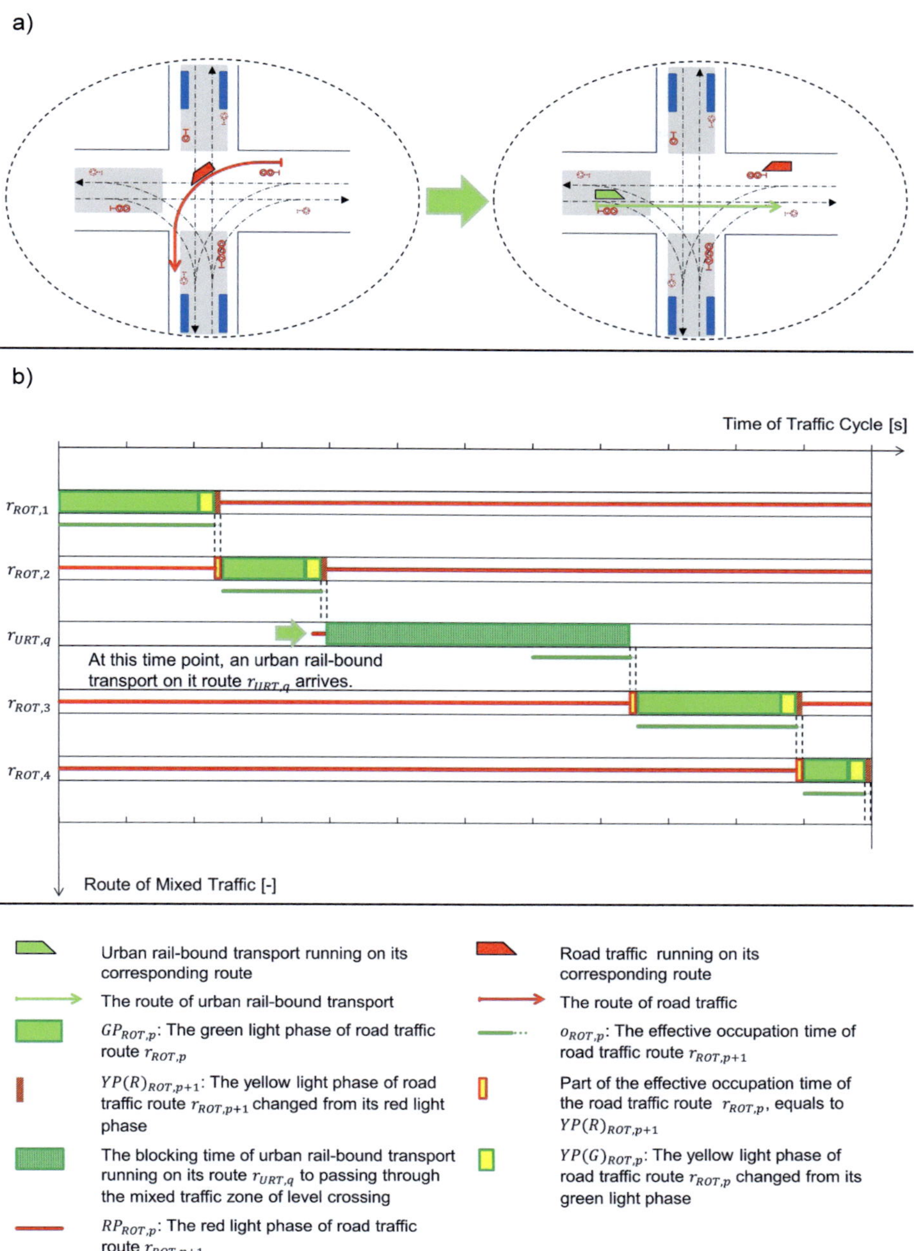

Figure 6-9: The Effective Occupation Time of a Road Traffic Route with Different Phases of a Traffic Control Signal

In Figure 6-9, the investigated example of the rotation of different phases of one road traffic route and between different road traffic routes in one traffic cycle is illustrated. It shows that a road traffic route $(r_{ROT,p})$ [17] can occupy the mixed traffic zone (level crossing) from the starting point in time of the green light phase $(GP_{ROT,p})$. After the green light phase finishes, then the yellow light phase changed from green light phase $(YP(G)_{ROT,p})$ starts, and is also part of the effective occupation time of the road traffic route $(o_{ROT,p})$.

For safety reasons, when the red light phase of this road traffic route is starting, the following green light phase of road traffic routes in the defined normal rotation cannot start immediately. A certain time of yellow light phase changed from red light phase $(YP(R)_{ROT,p+1})$ is necessary to ensure all the road traffic that occupied the level crossing during the previous green light phase $(GP_{ROT,p})$ and the yellow light phase $(YP(R)_{ROT,p})$ have left the level crossing. This is another part of the effective occupation time of the road traffic route $(o_{ROT,p})$, which is also a certain time duration of red light phase of the road traffic $(RP_{ROT,p})$ from the end of its yellow light phase $(YP(G)_{ROT,p})$ until the starting point of the next green light phase $(GP_{ROT,p+1})$ of the following road traffic route $(r_{ROT,p+1})$ in the defined normal rotation.

In conclusion, the effective occupation time of one road traffic route $r_{ROT,p}$ can be derived based on the defined traffic cycle.

$$o_{ROT,p} = GP_{ROT,p} + YP(G)_{ROT,p} + YP(R)_{ROT,p+1} \qquad\qquad (\text{6-25})$$

With:

$GP_{ROT,p}$ The green light phase of the road traffic route $r_{ROT,p}$.

$YP(G)_{ROT,p}$ The yellow light phase changed from green light phase the of road traffic route $r_{ROT,p}$.

$YP(R)_{ROT,p+1}$ The yellow light phase changed from red light phase of the next road traffic route $r_{ROT,p+1}$.

---

[17] The subscript "$p, p + 1$" here shows the example sequence of the green light phase of the road traffic routes in one traffic cycle. In other words, the green light phase of road traffic route $r_{ROT,p+1}$ is next to the green light phase of road traffic route $r_{ROT,p}$.

p The index of the road traffic routes, $p \in [1, m]$.

When there is urban rail-bound transport, the interfered road traffic route has to reduce its green light phase to its minimum green light phase ($(min)GP_{ROT,p}$) in some situations (see Chapter 6.4.4). Therefore, the minimum effective occupation time $(min)o_{ROT,p}$ of the road traffic route $r_{ROT,p}$ can be calculated.

The rotation of the traffic light phases controlled by the traffic control signal in one traffic cycle of the investigated example is shown in Figure 6-9. Table 6-5 shows the corresponding duration of the traffic light phases for the investigated example. The road traffic route $r_{ROT,1}$ is set to start at 0:00 with its green light phase $(GP_{ROT,1})$ with 34 seconds, and then is its yellow light phase $(YP(G)_{ROT,1})$ with 3 seconds. Considering the safety reasons, there is yellow light phase $(YP(G)_{ROT,2})$ with 1 second as the rest part of effective occupation time $(o_{ROT,1})$. Afterwards, the next road traffic route $r_{ROT,2}$ starts its green light phase $(GP_{ROT,2})$ with 20 seconds to occupy the investigated mixed traffic zone. The subsequent rotation is similar with corresponding time durations.

Route of Road Traffic	$GP_{ROT,p}$ [s]	$(min)GP_{ROT,p}$ [s]	$YP(G)_{ROT,p}$ [s]	$YP(R)_{ROT,p+1}$ [s]	$o_{ROT,p}$ [s]
$r_{ROT,1}$	34	14	3	1	38
$r_{ROT,2}$	20	14	3	1	24
$r_{ROT,3}$	35	14	3	1	39
$r_{ROT,4}$	10	11	3	1	14

Table 6-5: Time Duration of Traffic Light Phases of Road Traffic Routes for the Investigated Example of Level Crossing

When there is urban rail-bound transport, the normal rotation of road traffic routes is interrupted, which can be at a random point in time in the investigated time period. Figure 6-9 shows a possible case as an example, the urban rail-bound transport $r_{URT,q}$ that is exclusive to all of the four road traffic routes is arriving when the road traffic route $r_{ROT,2}$ is just on its yellow light phase $(YP(G)_{ROT,2})$. The urban rail-bound transport has to wait for a certain time period until the end of the occupation time of road traffic route $o_{ROT,2}$. Even though there is no "yellow light phase" for urban rail-bound transport before its progression, the road traffic route $r_{ROT,2}$ has to block the mixed traffic zone for a certain time period (with the time duration of $YP(R)_{ROT,3}$ in this investigated example) to make sure all the road traffic has left.

Calculation of Indicators

The **potential occupation time of road traffic** (O_{ROT}) for a whole traffic cycle can be derived through statistics on each effective occupation time of a road traffic route $o_{ROT,p}$ in one traffic cycle in the developed modeling of an event-driven system. Accordingly, the indicator of the potential occupation time of the road traffic (O_{ROT}) at an investigated mixed traffic zone in an investigated time period can be calculated. The value of the potential occupation time for the road traffic (O_{ROT}) is the maximal possible occupation time of road traffic for an investigated mixed traffic zone in an investigated time period. It is calculated as the sum of the effective occupation time of all exclusive road traffic routes in each traffic cycle in an investigated time period.

If there is no urban rail-bound transport, the road traffic occupies the level crossing all the time (except during the all red light phase) and the value of the potential occupation time for road traffic (O_{ROT}) should equal the investigated time period (T).

The time duration of an urban rail-bound transport hindered by the occupation of a road traffic route $r_{URT,p}$ is defined as the **hindrance time** $(h_{URT,q,p})$ of urban rail-bound transport route $r_{URT,q}$.

However, when one or more urban rail-bound transport vehicles exist, the hindrance time of the urban rail-bound transport caused by the road traffic route $r_{URT,p}$ in one traffic cycle can be derived by the hindrance times of the urban rail-bound transport vehicles on theirs routes. During one occupation time of a road traffic route $(o_{ROT,p})$, there can be one or more urban rail-bound transports hindered by the road traffic route $r_{ROT,p}$. The effective hindrance time of urban rail-bound transport caused by the road traffic route $r_{ROT,p}$ is calculated as the maximum hindrance time of all urban rail-bound transport hindered by the road traffic route $r_{ROT,p}$.

$$h_{URT,p} = max\left\{h_{URT,q,p}, h_{URT,q+1,p}, \cdots\right\}$$

(6-26)

With:

$h_{URT,p}$	The effective hindrance time of a urban rail-bound transport route caused by road traffic route $r_{ROT,p}$ in one traffic cycle at an investigated mixed traffic zone.
$h_{URT,q,p}$	The hindrance time of an urban rail-bound transport route $r_{URT,q}$ caused by road traffic route $r_{ROT,p}$.
q	The index of the urban rail-bound transport routes, $q \in [1, n]$.
p	The index of the road traffic routes, $p \in [1, m_e]$.

Accordingly, the other indicator of the **hindrance time of urban rail-bound transport caused by road traffic (H_{URT})** at an investigated mixed traffic zone in an investigated time period can be calculated. The value of the hindrance time of the urban rail-bound transport caused by road traffic (H_{URT}) is the total occupation time of road traffic, at which the urban rail-bound transport vehicles are hindered by the occupation of the road traffic at the investigated mixed traffic zone in an investigated time period.

Based on the pre-defined locking table of route related locking for urban mixed traffic and the relevant rules for the investigated mixed traffic zone with basic information [infrastructure, operating program (specific scheduled timetable) of the urban rail-bound transport and urban roadways, and the normal rotation of traffic light phases of road traffic route], the interactions between the road traffic and urban rail-bound transport at a mixed traffic zone level crossing can be modeled in an established event-driven system. There are various cases that can be triggered under different conditions, which are described in Chapter 6.4.4.

Extension of Algorithm for Shared Roads

For the mixed traffic zone of shared roads, the model with the event-driven system has to be further developed based on the model for level crossings. The shared road can be treated as a series of fictitious adjacent level crossings step by step at the possible location points (LP) along the shared road from the start point s_4. For example, to the end point s_9 of one route (r_{URT}: $s_4 \rightarrow s_9$ or r_{ROT}: $s_4 \rightarrow s_9$) as described previously (see Figure 6-7).

The urban rail-bound transport on the route $r_{URT}: s_4 \rightarrow s_9$ may be influenced randomly at any possible location point (LP) along the shared road, which is decided by the road traffic on the route $r_{ROT}: s_4 \rightarrow s_9$ in one compatible route combination (CRC: $s_4 \rightarrow s_9$) moving along the shared road ahead of it ($r_{URT}: s_4 \rightarrow s_9$). When the traffic flow of the road traffic on the route $r_{ROT}: s_4 \rightarrow s_9$ on the shared road is very high, the urban rail-bound transport on the route $r_{URT}: s_4 \rightarrow s_9$ may be blocked by the first possible location point (LP_1) after the start point (s_4), because the compatible routes of combination ($CRC: s_4 \rightarrow s_9$) along the shared road are controlled and locked by the traffic control signal of the following adjacent level crossing at the end of the shared road.

It is possible for urban rail-bound transport to operate from the start point s_4 along the shared road, if the road traffic ahead of it is not congested. When the urban rail-bound transport on the route $r_{URT}: s_4 \rightarrow s_9$ is blocked on the shared road at a location point, it means that the section of shared road before this urban rail-bound transport is full with other mixed traffic, and the traffic control signal of the following adjacent level crossing at the end point of the route s_9 has a red or yellow light rather than a green light. When the effective occupation times of the other routes that are exclusive to the corresponding connected route start with s_9 at the following adjacent level crossing are over, it is then free for both the urban rail-bound transport $r_{URT}: s_4 \rightarrow s_9$ and the road traffic $r_{ROT}: s_4 \rightarrow s_9$. They can move again for the time duration of the green light phase of the corresponding connected route that starts with s_9. Some of road traffic in front of the urban mixed-rail bound transport can leave the shared road, and then the rest of them will be blocked again.

In this algorithm on the shared road, when the urban rail-bound transport is not blocked by road traffic in front of it, it runs at its own determined velocity. When it is hindered by road traffic, it runs with the road traffic at a lower speed. The different driving dynamics including acceleration and deceleration between the train and the road traffic are not considered because the running time of urban rail-bound transport is restricted by the movements of the road traffic. From the view of running time, there are no obvious differences.

The series of fictitious adjacent level crossings with traffic control signals modeled on the shared roads are used to represent the possible location points (LP) at which the urban rail-bound transport is hindered by the road traffic. Therefore, when the urban rail-bound transport is hindered randomly by road traffic, it may occur at any fictitious adjacent level crossing (i.e. any possible location points). Once the hindrance occurs, the movement of the urban rail-bound transport is controlled by the rotation of the traffic control signal of the corresponding connected route starting at s_9.

In conclusion, the influences of the road traffic can be derived by the hindrance time of the urban rail-bound transport from the location point at which the urban rail-bound transport is hindered by the road traffic along the shared road before it to the end point of the shared road. Therefore, through the indication of the red light phase of the traffic control signals at the series of fictitious adjacent level crossings, the location point (LP) at which the urban rail-bound transport is hindered by road traffic can be determined. Therefore, the number of times that the urban rail-bound transport has to stop and wait for the road traffic in front of it receiving the next green light phase of the corresponding traffic control signal starting at s_9 can be calculated as:

$$h = \left\lfloor \frac{S - l}{GP \cdot v} \right\rfloor + 1 \qquad\qquad (\text{6-27})$$

With:

h The number of times that the urban rail-bound transport is hindered along the shared road.

S The length of the shared road, for example, the length from the start point s_4 to the end point s_9.

l The distance from the location point (LP), at which the urban rail-bound transport is first hindered by the road traffic to the end point of the shared road.

GP The time duration of the green light phase of the traffic control signal of the corresponding connected route at the following adjacent level crossing start with s_9.

v The velocity of the urban mixed traffic (road traffic).

Therefore, the indicator of the hindrance time of the urban rail-bound transport caused by the road traffic (H_{URT}) at a shared road in an investigated time period can be calculated as the sum of the hindrance times of each urban rail-bound transport vehicle caused by road traffic along the shared road.

Through the model of the automatic event-driven system, a large sample of simulation models with random event triggers can be generated without accruing high costs. Accordingly, a large amount of results of the potential occupation time for the road traffic (O_{ROT}) and the hindrance time of the urban rail-bound transport caused by the road traffic (H_{URT}) at an investigated mixed traffic zone during an investigated time period can be calculated. It is plausible that the average of all of the results can, to a certain extent, show the possible road traffic influences in reality.

6.4.3 Determination of the Parameters

The two important parameters of throughput capacity with road traffic influences ($\widehat{DS\,LF}$) and the starting waiting time with road traffic influences (d) have to be estimated to derive the waiting time function using this developed algorithm.

For a mixed traffic zone level crossing, the two parameters can be estimated from the two calculated indicators with the modeling of event-driven system: the potential occupation time for the road traffic (O_{ROT}) and the hindrance time of the urban rail-bound transport caused by the road traffic (H_{URT}).

For the mixed traffic zone of shared roads, the relationship between the urban rail-bound transport and road traffic at the investigated area is codependent rather than competitive along the shared road, even though the urban rail-bound transport is hindered by the road traffic in front of it. Therefore, the parameter of the starting waiting time with road traffic influences (d) can be estimated with the calculated indicator of the hindrance time of the urban rail-bound transport caused by road traffic (H_{URT}). The other parameter of throughput capacity with road traffic influences ($\widehat{DS\,LF}$) can be calculated based on the algorithm proposed in [Schmidt 2009] and developed in [Chu 2014] for determining the throughput capacity in railway systems through comparison of the number of trains entering and exiting the investigated area during the investigated time period with the stepwise-varied traffic flows of trains.

In Figure 6-10 the relationship between the y-axes with the value of the potential occupation time for the road traffic (O_{ROT}) and the hindrance time of the urban rail-bound transport caused by the road traffic (H_{URT}) as well as the x-axis with the stepwise-varied traffic flows of urban rail-bound transport are illustrated. Due to the competitive relationship between the urban rail-bound transport and the road traffic at the investigated level crossing, the higher the traffic flow of the urban rail-bound transport is, the greater the number of trains being hindered by the road traffic is. Meanwhile the value of the hindrance time of the urban rail-bound transport caused by the road traffic (H_{URT}) will be increased. Respectively, the potential occupation time of the road traffic (O_{ROT}) is decreased due to the increased occupancy of the urban rail-bound transport at the mixed traffic zone level crossings.

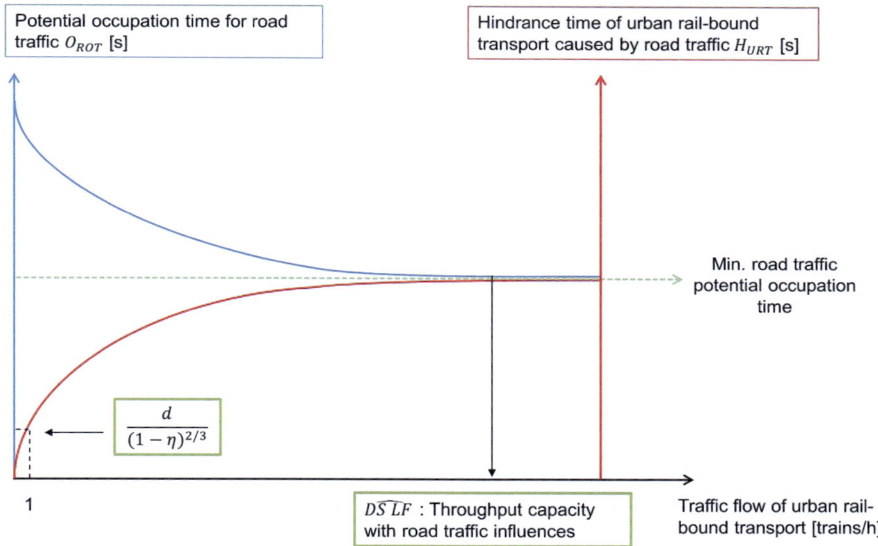

Figure 6-10: Relationship between potential Occupation Time for Road Traffic (O_{ROT}) and the Hindrance Time of the Urban Rail-bound Transport by the Road Traffic (H_{URT}) (Source: modified based on [Martin & Di Liu 2016])

According to the basic concept of the relationship between the potential occupation time for the road traffic (O_{ROT}) and the hindrance time of the urban rail-bound transport by the road traffic (H_{URT}), the competition between them is restricted by the pre-defined rules. Even though the urban rail-bound transport has a higher priority to occupy the mixed traffic zone in general, the rule of limitations of traffic light phase for road traffic has to be complied with.

Therefore, theoretically it can reach the minimum potential occupation time for the road traffic, which means that each of the road traffic routes ($r_{ROT,p}$) at the investigated mixed traffic zone level crossing can only occupy the level crossing with the minimum effective occupation time $(min)o_{ROT,p}$ of a road traffic route $r_{ROT,p}$. However, this is a theoretical value. If the road traffic route is compatible with an ongoing urban rail-bound transport route, its green light phase cannot drop to minimum and even can be extended in practice. Besides, the investigated level crossing can be occupied by each of the road traffic routes when it is already waiting for the red light phase of the traffic control signal for the limitation duration during the investigated time period in this situation. This implies that the urban rail-bound transport can occupy the investigated level crossing at all other times in the investigated time period.

Once the road traffic can occupy the level crossing, the potential occupation time will be reduced to the minimum potential occupation time and the traffic flow of the urban rail-bound transport increases up to the throughput capacity with road traffic influences ($\widehat{DS\,LF}$). Meanwhile, a convergence will be reached between the potential occupation time for the road traffic (O_{ROT}) and the hindrance time of the urban rail-bound transport caused by the road traffic (H_{URT}). At this point, the occupation of the road traffic will always hinder the urban rail-bound transport. Therefore, once the relation of the value of the potential occupation time for the road traffic (O_{ROT}) and the hindrance time of the urban rail-bound transport caused by the road traffic (H_{URT}) is derived, the value of the throughput capacity with road traffic influences ($\widehat{DS\,LF}$) can be obtained.

With regard to the mixed traffic zone of shared roads, the throughput capacity with road traffic influences ($\widehat{DS\,LF}$) can be determined using the number of entering and exiting urban rail-bound transport vehicles with stepwise-varied traffic flows recorded by the protocol from the event-driven system modeling by the developed algorithm. (see [Chu 2014])

With the curve of the hindrance time of urban rail-bound transport by road traffic (H_{URT}) at a investigated mixed traffic zone of level crossing or shared road (see Figure 6-10), the starting point of the waiting time function (6-11) ($d = d_5/(1 - \eta)^{2/3}$) can also be determined. According to the improved waiting time function (6-11) based on the existing one (6-5), the parameter d shows the waiting time when the number of urban rail-bound transport vehicles is in the range of (0, 1] from the statistical point of view. It indicates the probability of the occurred waiting time due to the operational hindrance by the influences of the road traffic when there is only one train of urban rail-bound transport (U_i). The parameter d_{U_i} for the waiting time function (6-11) equals the value of the hindrance time of the urban rail-bound transport U_i caused by the road traffic (H_{URT}) when the number of urban rail-bound transport vehicles equals 1.

Hence the relation of the value of the potential occupation time for the road traffic (O_{ROT}) and the hindrance time of the urban rail-bound transport caused by the road traffic (H_{URT}) for the mixed traffic zone level crossing is derived, and the parameters of the throughput capacity with road traffic influences $(\widehat{DS\,LF})$ and the starting point of the waiting time function (d) for the adapted waiting time function (6-11) can be determined. Similarly, the hindrance time of the urban rail-bound transport caused by road traffic (H_{URT}) for the mixed traffic zone of shared roads is available to derive the parameter of the starting point of the waiting time function (d) for the shared roads. Moreover, the number of entering and exiting trains (urban rail-bound transports) with stepwise-varied traffic flows are also recorded through the model of event-driven system and further used to determine the parameter of the throughput capacity with road traffic influences $(\widehat{DS\,LF})$ of the mixed traffic zone of shared roads.

6.4.4   Event-driven Cases

In this algorithm, the modeling of random interactions between the urban rail-bound transport and the road traffic is built with an event-driven system. The model is created to describe the responses with the trigger of predicted interactions between the urban rail-bound transport and the road traffic in the form of an "event". Those responses can represent the randomness of an occurrence with possible status changes to the mixed traffic at a specific point in time during the investigated time period.

Various cases are defined here as the various responses (the predicted interactions) of the urban rail-bound transport and road traffic when an event is triggered. In this model, once an urban rail-bound transport vehicle arrives at the investigated mixed traffic zone at various specific points in time during the investigated time period, an event is triggered. In the simulation process with the model, the possible occurrence of an event is random. It is based on the assessment and judgement of the specific situation at the specific point in time to select the corresponding case. Therefore, the algorithm can be implemented in the following steps with various event triggers under various conditions in the workflow shown in Figure 6-11.

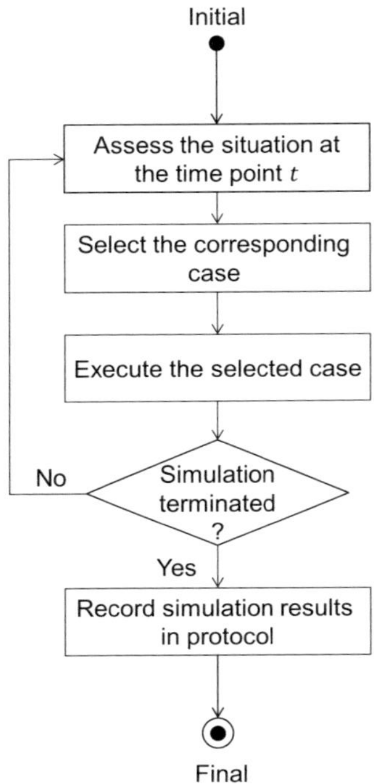

Figure 6-11: The Workflow of the Developed Algorithm with Event-Driven Simulation

<u>Step 1</u>: Assess the situation at the point in time t

Assess the temporal situation and status of the interactions between the urban rail-bound transport and the road traffic when an event occurs at a specific point in time during the investigated time period. When an urban rail-bound transport vehicle arrives at the investigated mixed traffic zone at a specific point in time, it is possible that the road traffic will influence the following operation of the urban rail-bound transport. Accordingly, it is necessary to judge the operation status changes of the urban rail-bound transport and the road traffic starting from this point in time.

<u>Step 2</u>: Select the corresponding case

According to the assessed status of the interaction in Step 1, the corresponding response is selected from the determined cases. The event is triggered and the selection is dependent on the temporal status. The corresponding case is selected in terms of the triggered event at the point in time.

<u>Step 3</u>: Execute the selected case

The responses of the urban rail-bound transport and the road traffic depicted in the selected case are executed. As a result, their operation status is changed in the simulation process. The road traffic influences are represented through the responses in the model. In addition, the subsequent operations of urban rail-bound transport and road traffic are also defined in each of the cases, when the changes caused by the triggered event are completed.

<u>Step 4</u>: Judge whether the simulation is terminated

When the urban mixed traffic completes its responses and changes of status due to the triggered event, its subsequent operations are in operation. It is necessary to judge whether the simulation has been terminated or not based on the time in the investigated time period. When the investigated time period is over, the simulation will be terminated. Afterwards, the results of the simulation are recorded in the output protocol. If the simulation is still in progress, the iterative process is still running and prepared to assess the situation again. When there is another event triggered in the following progress, the four steps will be reiterated to be implemented until the terminated condition of the simulation is achieved at the end of the investigated time period.

According to the inputs and relevant rules described in Subchapter 6.4.2, for mixed traffic zone level crossings, there are six various cases that will be selected and executed in Step 2 and Step 3 under various situations and statuses of the interaction between the urban rail-bound transport and the road traffic. The selected case will be conducted when an event is triggered at a random specific point in time t during the investigated time period.

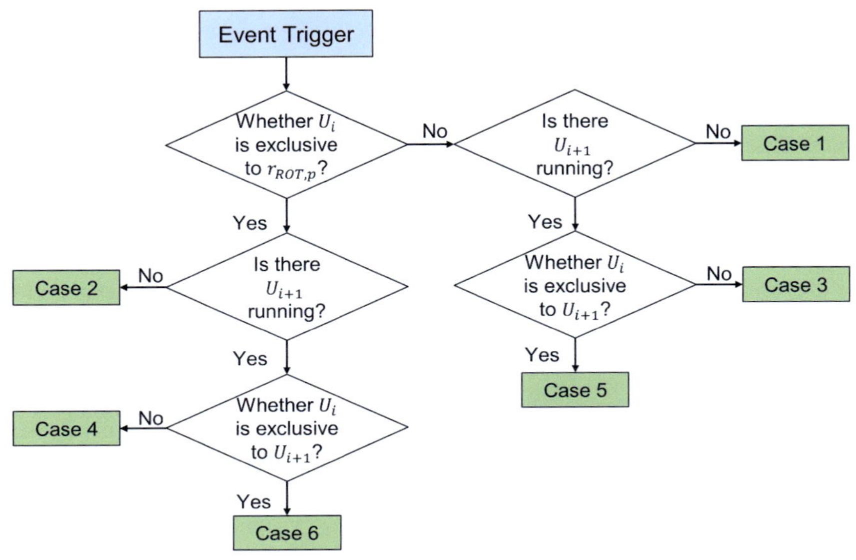

U_i	The urban rail-bound transport on its route $r_{URT,q}$ arrives at an investigated level crossing.
U_{i+1}	The urban rail-bound transport on its route $r_{URT,q+1}$ occupies the investigated level crossing when the urban rail-bound transport on its route $r_{URT,q}$ arrives
$r_{ROT,p}$	The road traffic on its route $r_{ROT,p}$ occupies the investigated level crossing when the urban rail-bound transport on its route $r_{URT,q}$ arrives

Figure 6-12: Event Trigger Conditions of Six Cases for Mixed Traffic Zone of Level Crossing

It is possible that the occurred case varies with the different specific points in time when the event is triggered because at different points in time, the occupation statuses of the road traffic at the level crossings are different. The conditions that trigger an event for the six cases are illustrated in Figure 6-12, when an urban rail-bound transport vehicle U_i on its route $r_{URT,q}$ arrives at an investigated level crossing.

<u>Case 1</u>: With compatible road traffic route $r_{ROT,p}$, without other urban rail-bound transport.

When an urban rail-bound transport vehicle U_i on its route $r_{URT,q}$ arrives at an investigated mixed traffic zone level crossing, the road traffic route $r_{ROT,p}$ compatible with the urban rail-bound transport vehicle U_i is running at the same moment; and there is no other urban rail-bound transport operating at the mixed traffic zone level crossing. In this situation, the urban rail-bound transport vehicle U_i on its route $r_{URT,q}$ arrived and is free to immediately enter and occupy the level crossing without extra waiting time.

a) If the surplus time duration of the effective occupation time $o_{ROT,p}$ of the road traffic route $r_{ROT,p}$ is longer than the blocking time needed to pass through the level crossing of the urban rail-bound transport vehicle U_i, there are no road traffic influences on the urban rail-bound transport U_i. Additionally, the normal rotation of the traffic control signals for the road traffic is not interrupted.

b) On the contrary, if the blocking time of the urban rail-bound transport vehicle U_i is longer than the surplus time duration of the effective occupation time $o_{ROT,p}$, of the road traffic route $r_{ROT,p}$, which means the occupation of road traffic route $r_{ROT,p}$ is scheduled to finish earlier, the urban rail-bound transport vehicle U_i is still running at the level crossing at that time. It is necessary to judge whether there are other road traffic routes compatible with the urban rail-bound transport vehicle U_i.

If all of the other road traffic routes are exclusive to the urban rail-bound transport vehicle U_i, the normal rotation of the traffic control signals for the road traffic is interrupted until the urban rail-bound transport U_i leaves the level crossing without road traffic influences. In this situation, the road traffic route $r_{ROT,p}$ can have its occupation time extended to the exit time of the urban rail-bound transport vehicle U_i.

If there are one or more other road traffic routes compatible with the urban rail-bound transport vehicle U_i, according to the pre-defined sequential rule, it can be determined which one of the compatible road traffic routes is allowed to proceed to enter the mixed traffic zone. In this case, there are one or more compatible road traffic routes allowed to enter the level crossing, which means they can start to occupy the level crossing with the urban rail-bound transport U_i at the same time. Here, the temporal status of the interactions between the urban rail-bound transport and the road traffic is iterative to the state of the urban rail-bound transport U_i entering the level crossing in Case 1.

<u>Case 2</u>: With exclusive road traffic route $r_{ROT,p}$, without other urban rail-bound transport.

In this case, except for the same state of Case 1: without other urban rail-bound transport. The temporal status is different when an urban rail-bound transport vehicle U_i on its route $r_{URT,q}$ arrives at an investigated mixed traffic zone level crossing; the occupying road traffic route $r_{ROT,p}$ is exclusive to it.

When the urban rail-bound transport U_i asks for permission to pass through, if the road traffic route $r_{ROT,p}$ hasn't yet completed its minimum green light phase $((min)GP_{ROT,p})$, it has to continue to operate until the minimum green light phase has been reached. And then the road traffic route can proceed to start its yellow light phase $(YP(G)_{ROT,p})$, which prepares to give the priority to the urban rail-bound transport vehicle U_i. Or else, the road traffic route $r_{ROT,p}$ has already gone over its minimum green light phase $((min)GP_{ROT,p})$, and it has to proceed to start its yellow light phase $(YP(G)_{ROT,p})$ at the same moment that the urban rail-bound transport vehicle U_i arrives at the level crossing.

As a result of both of the above probabilities, the operation of the road traffic route $r_{ROT,p}$ is interrupted and first finishes its occupation time at the investigated level crossing. Afterwards, the urban rail-bound transport vehicle U_i will have the right to enter the level crossing. As long as there are no other road traffic routes already waiting for the green light phase and the limitation of the maximum red light phase has been exceed, the urban rail-bound transport U_i has a higher priority for passing through the level crossing. Therefore, the occupied road traffic route $r_{ROT,p}$ has to reduce its green light phase $(GP_{ROT,p})$ until completing its minimum green light phase $((min)GP_{ROT,p})$. The urban rail-bound transport vehicle U_i only has to complete its occupancy at the start point of the level crossing.

Meanwhile, it needs to be assessed whether there is another road traffic route that is compatible with the urban rail-bound transport vehicle U_i that can start its green light phase at the same time based on the sequential rule. If no other compatible road traffic route exists, which means all of the other road traffic routes are exclusive to the urban rail-bound transport vehicle U_i, it can pass through the level crossing. Otherwise, the urban rail-bound transport vehicle U_i and the compatible road traffic route can start their green light phase to pass through the level crossing together at the same time. This is the same temporal status as the initial status of Case 1.

Once there is another road traffic route waiting for the green light phase and is over the limitation of the maximum red light phase, it is necessary to assess whether this road traffic route is compatible with the urban rail-bound transport vehicle U_i. If not, the urban rail-bound transport vehicle U_i cannot gain permission to pass through the level crossing after the road traffic route $r_{ROT,p}$. It has to continue to wait at the start point of the level crossing, and the temporal status of the interactions between the urban rail-bound transport and the road traffic is iterated back to the initial state of Case 2. The iterative execution process of Case 2 has to be carried out again. Conversely, the urban rail-bound transport vehicle U_i can proceed to enter the level crossing with the road traffic route that has exceeded the maximum red light phase, and they can occupy the level crossing at the same time.

<u>Case 3</u>: With compatible road traffic route $r_{ROT,p}$ and other urban rail-bound transport U_{i+1}[18] on its route $r_{URT,q+1}$.

The temporal status of the triggered event for the execution of Case 3 is: both the road traffic route $r_{ROT,p}$ and another urban rail-bound transport U_{i+1} on its route $r_{URT,q+1}$ are compatible with the urban rail-bound transport vehicle U_i that arrived. The urban rail-bound transport vehicle U_i can proceed to enter the level crossing without interruption. The state can be regarded as the urban rail-bound transport vehicle U_i arrives at the level crossing in the process of Case 1.

By now, the road traffic route $r_{ROT,p}$ needs some time to complete its effective occupation time (in normal rotation), which is defined as the surplus effective occupation time for the road traffic route $r_{ROT,p}$. Meanwhile, the urban rail-bound transport vehicle U_{i+1} that is passing through the level crossing also needs surplus time added to its residual blocking time at the level crossing. The urban rail-bound transport vehicle U_i needs the entire blocking time to pass through the level crossing.

---

[18] There is no sequence meaning in timetable among the urban rail-bound transport U_i, U_{i+1}, etc., which means the possible urban rail-bound transport on its corresponding routes $r_{URT,q}$, $r_{URT,q+1}$, etc. also without meaning of sequence of normal rotation in traffic cycle.

a) If the surplus effective occupation time for the road traffic route $r_{ROT,p}$ lasts longer than both of the (surplus) blocking times of the urban rail-bound transport vehicles U_i and U_{i+1}, both of them will leave the level crossing before the road traffic route $r_{ROT,p}$ completes its occupancy. When the urban rail-bound transport vehicle U_i leaves earlier than the urban rail-bound transport vehicle U_{i+1}, the urban rail-bound transport vehicle U_{i+1} can also leave the level crossing without being affected by the road traffic influences. The road traffic operates normally based on the normal rotation.

In the case where the surplus blocking time of the urban rail-bound transport vehicle U_{i+1} is shorter than the blocking time of the urban rail-bound transport vehicle U_i, it will leave the level crossing first. At the moment, It is necessary to judge whether there are other road traffic routes compatible with both the urban rail-bound transport vehicle U_i and the road traffic route $r_{ROT,p}$. Similar to the assessment in Case 1 and Case 2, and according to the sequential and limitation rules, if no other road traffic routes meet the judging conditions, this case is completed. The road traffic continues to operate normally.

Conversely, a road traffic route compatible with both of the urban rail-bound transport vehicle U_i and the road traffic route $r_{ROT,p}$ can proceed to start its green light phase. Because all of the road traffic routes occupying the level crossing at this time are compatible with each other and with the urban rail-bound transport vehicle U_i, this temporal status can be regarded as one of the states in the process of Case 1. It is the iterative process of the state of the urban rail-bound transport vehicle U_i entering the level crossing in Case 1.

b) If the surplus effective occupation time for the road traffic route $r_{ROT,p}$ lasts for a shorter period of time than both of the (surplus) blocking times of the urban rail-bound transport vehicles U_i and U_{i+1}, the road traffic route $r_{ROT,p}$ completes its occupancy first, and then there are only the two urban rail-bound transport U_i and U_{i+1} in the level crossing. Likewise, on the basis of the sequential and limitation rules, it is necessary to judge whether or not other road traffic routes are compatible with them. If not, it is possible to extend the occupation time of the road traffic route $r_{ROT,p}$. When the blocking time of one of the two urban rail-bound transport vehicles is over earlier, there is only one urban rail-bound transport vehicle in the level crossing and the surplus effective occupation time of the road traffic route $r_{ROT,p}$ is shorter than the blocking time. This temporal status is also able to be regarded as one of the states in the process of Case 1 b).

Otherwise, a road traffic route is compatible with the two urban rail-bound transport vehicles U_i and U_{i+1}, which proceeds to start its green light phase. This is the iterative process of the state of the urban rail-bound transport vehicle U_i entering the level crossing in Case 3.

c) In this case, the surplus effective occupation time for the road traffic route $r_{ROT,p}$ is longer than the minimum (surplus) blocking time of the two urban rail-bound transport vehicles U_i and U_{i+1}, and shorter than their maximum (surplus) blocking time. Because there are no other road traffic route compatible with both of the road traffic route $r_{ROT,p}$ and the urban rail-bound transport vehicle U_{i+1}, when the urban rail-bound transport vehicle U_i leaves earlier than the urban rail-bound transport vehicle U_{i+1}, the urban rail-bound transport vehicle U_{i+1} still continues into the level crossing after the road traffic route $r_{ROT,p}$ completes its surplus effective occupation time. This temporal status is also able to be regarded as one of the states in the process of Case 1 b).

If the blocking time of the urban rail-bound transport vehicle U_i is longer than the surplus blocking time of the urban rail-bound transport vehicle U_{i+1}, it is necessary to judge whether there is other road traffic compatible with both the road traffic route $r_{ROT,p}$ and the urban rail-bound transport U_i. If not, the situation is the same as when the urban rail-bound transport vehicle U_i leaves earlier than the urban rail-bound transport vehicle U_{i+1}.

Conversely, the road traffic compatible with both the road traffic route $r_{ROT,p}$ and the urban rail-bound transport vehicle U_i is allowed to proceed to start its green light phase. Then, if it completes its occupancy earlier than the road traffic route $r_{ROT,p}$, the occupation time of it and the road traffic route $r_{ROT,p}$ can be extended to the end of the blocking time of the urban rail-bound transport vehicle U_i. Afterwards, the road traffic will continue to operate normally. If the road traffic route $r_{ROT,p}$ completes its occupancy first, the urban rail-bound transport vehicle U_i and the other road traffic route are still running in the level crossing. This is the same situation as Case 1.

<u>Case 4</u>: With exclusive road traffic route $r_{ROT,p}$, simultaneously with other compatible urban rail-bound transport vehicle U_{i+1}

The event to execute Case 4 is triggered when an urban rail-bound transport vehicle U_i on its route $r_{URT,q}$ arrives at the investigated level crossing, and the road traffic route $r_{ROT,p}$ is exclusive to the urban rail-bound transport vehicle that arrived on its route $r_{URT,q}$ that is compatible with the urban rail-bound transport vehicle U_{i+1} on its route $r_{URT,q+1}$.

This case is the extension of the basis of Case 2. The operation of the road traffic route $r_{ROT,p}$ is interrupted by the urban rail-bound transport vehicle U_i. The reaction of the road traffic route $r_{ROT,p}$ is the same as in the description in Case 2. It has to first complete its effective occupation time according to the specific conditions, and then the urban rail-bound transport vehicle U_i is allowed to enter the level crossing. However, there are some differences from Case 2. It is necessary to judge which of the surplus occupation times of the road traffic route $r_{ROT,p}$ and the surplus blocking time of urban rail-bound transport vehicle U_{i+1} is longer.

a) If the road traffic route $r_{ROT,p}$ completes its effective occupation time earlier. it is possible that there are other road traffic routes compatible with the two urban rail-bound transports based on the sequential and limitation rules. The compatible road traffic route is allowed to proceed to start its green light phase. At this moment, the status is able to be regarded as the state of the urban rail-bound transport vehicle U_i entering the level crossing in Case 3. Two urban rail-bound transport vehicles U_i and U_{i+1}, and also a compatible road traffic route occupy the level crossing at the same time.

If no other road traffic routes are compatible with the two urban rail-bound transports, only two urban rail-bound transport vehicles U_i and U_{i+1} operate at the level crossing until one of them leaves. Afterwards, it is necessary to judge whether there are other road traffic routes compatible with the one still running at the level crossing. If not, according to the sequential and limitation rules, the urban rail-bound transport continues to operate until leaving the level crossing. The case is over; the road traffic starts to continue its normal rotation again. Otherwise, the road traffic route compatible with the urban rail-bound transport vehicle that is still running is allowed to proceed to start its green light phase. This temporal status is regarded as the state of the urban rail-bound transport vehicle U_i entering the level crossing in Case 1.

b) If the urban rail-bound transport vehicle U_{i+1} leaves the level crossing before the road traffic $r_{ROT,p}$ completes its effective occupation time, this is precisely the status of the initial state of Case 2 when the blocking time of the urban rail-bound transport vehicle U_{i+1} is over.

<u>Case 5</u>: With compatible road traffic route $r_{ROT,p}$, simultaneously with other exclusive urban rail-bound transport vehicle U_{i+1} on its route $r_{URT,q+1}$.

The initial state of this case is contra to that of Case 4. The road traffic route $r_{ROT,p}$ is compatible with the urban rail-bound transport vehicle U_i on its route $r_{URT,q}$ that is exclusive to the urban rail-bound transport vehicle U_{i+1}. Due to the exclusive urban rail-bound transport vehicle U_{i+1}, the urban rail-bound transport vehicle U_i cannot directly enter the level crossing. It has to wait until the exclusive urban rail-bound transport vehicle U_{i+1} leaves the level crossing.

a) It is possible that the surplus effective occupation time of the road traffic route $r_{ROT,p}$ is longer than the surplus blocking time of the urban rail-bound transport vehicle U_{i+1}. This happens when the road traffic route $r_{ROT,p}$ is compatible with the urban rail-bound transport vehicle U_i, the temporal status is the same as the state of the urban rail-bound transport U_i entering the level crossing in the process of Case 1.

b) Conversely, the surplus effective occupation time of the road traffic route $r_{ROT,p}$ is completed before the urban rail-bound transport vehicle U_{i+1} leaves the level crossing. When the road traffic route $r_{ROT,p}$ completes its occupation time, it is necessary to judge whether there is other road traffic compatible with the urban rail-bound transport vehicle U_{i+1} according to the sequential and limitation rules.

If not, the urban rail-bound transport vehicle U_{i+1} continues its operation with the road traffic route $r_{ROT,p}$ extending its occupancy until the end of the blocking time of the urban rail-bound transport vehicle U_{i+1}. Afterwards the urban rail-bound transport vehicle U_i proceeds to enter the level crossing. It has to be judged whether there is other road traffic compatible with the urban rail-bound transport vehicle U_i has to be carried out. If not, the urban rail-bound transport vehicle U_i operates normally until leaving; the case is terminated. The road traffic follows the normal rotation again. Otherwise, the compatible road traffic route is able to proceed to start its green light phase at the same time. At this moment, the status of the urban rail-bound transport vehicle U_i entering the level crossing is the same as in the process of Case 1.

Otherwise, other road traffic route compatible with the urban rail-bound transport vehicle U_{i+1} is allowed to proceed to start its green light phase. The urban rail-bound transport vehicle U_i is still waiting at the start point of the level crossing. Through judgement of the compatibility between the road traffic route and the urban rail-bound transport vehicles U_{i+1} and U_i. If they are compatible, the temporal status triggers to the initial state of this Case 5 iteratively. On the contrary, if they are exclusive to each other, the temporal status is iterative to the initial state of Case 6.

<u>Case 6</u>: With exclusive road traffic route $r_{ROT,p}$ and other urban rail-bound transport vehicle U_{i+1} on its route $r_{URT,q+1}$.

This case is also expanded from Case 2, but different from Case 4. The road traffic route $r_{ROT,p}$ and other urban rail-bound transport vehicle U_{i+1} are both exclusive to the arrived urban rail-bound transport vehicle U_i on its route $r_{URT,q}$. For the exclusive road traffic route $r_{ROT,p}$, it has to complete its minimum effective occupation time as well, which refers to the description in Case 2 and 4; the urban rail-bound transport U_i can proceed to enter the level crossing. The same as the judgement in Case 4, comparison of the surplus minimum effective occupation time of road traffic route $r_{ROT,p}$ and the surplus blocking time of urban rail-bound transport vehicle U_{i+1}.

    a) If the urban rail-bound transport vehicle U_{i+1} leaves the level crossing in advance, it operates normally to pass through the level crossing. At this moment, the urban rail-bound transport vehicle U_i is still waiting at the start point of the level crossing. Meanwhile there is a road traffic route $r_{ROT,p}$ exclusive to the urban rail-bound transport vehicle U_i. This temporal status can be regarded as the initial state of Case 2.

b) On the contrary, the surplus minimum effective occupation time of the road traffic route $r_{ROT,p}$ can be shorter than the surplus blocking time of the urban rail-bound transport vehicle U_{i+1}. Because the urban rail-bound transport vehicle U_{i+1} is also exclusive to the urban rail-bound transport vehicle U_i, U_i cannot enter the level crossing unless the urban rail-bound transport vehicle U_{i+1} has already left. The road traffic route $r_{ROT,p}$ is considered to extend its minimum effective occupation time.

If the surplus blocking time of the urban rail-bound transport vehicle U_{i+1} is shorter than the surplus effective occupation time of the road traffic route $r_{ROT,p}$, the road traffic route $r_{ROT,p}$ can extend its minimum effective occupation time to the end of the blocking time of the urban rail-bound transport vehicle U_{i+1}. Both of them finish theirs occupancy at the same time.

And then, according to the limitation rule, it is possible to judge whether the urban rail-bound transport vehicle U_i has the right to proceed to enter the level crossing. If not, it has to continue to wait at the start point of the level crossing. Otherwise, it is able to proceed to enter the level crossing. It is also needed to be judged whether there are other road traffic routes compatible with the urban rail-bound transport vehicle U_i. If not, it operates to pass through the level crossing and the road traffic follows their normal rotation. Conversely, the compatible road traffic route is allowed to start its green light phase together with the urban rail-bound transport vehicle U_i, which is the temporal status of U_i entering the level crossing in the process of Case 1.

Conversely, the surplus effective occupation time of the road traffic route $r_{ROT,p}$ can be shorter than the surplus blocking time of urban rail-bound transport vehicle U_{i+1}. At this moment, the urban rail-bound transport vehicle U_i is still waiting at the start point of the level crossing. The effective occupation time of the road traffic route $r_{ROT,p}$ will be completed before or extended to the end of the blocking time of the urban rail-bound transport vehicle U_{i+1}. It is necessary to judge whether there is another road traffic route exceeding its maximum red light phase. If not, the urban rail-bound transport U_i can proceed to enter the level crossing and it will be judged whether there is other road traffic that can also start its green light phase based on the relevant rules. Otherwise, the urban rail-bound transport vehicle U_i cannot enter the level crossing with the exclusive road traffic route exceeding its maximum red light phase. This temporal status can be regarded as the state in the process of Case 2.

There is one case for the mixed traffic zone of shared roads. The event is triggered when the urban rail-bound transport vehicle U_i is hindered by road traffic at a lower speed for the first time along the investigated shared road at a random specific point in time t during the investigated time period. As described in Subchapter 6.4.2, once the urban rail-bound transport vehicle U_i is hindered and must stop due to the road traffic the first time at a random location point at a fictitious adjacent level crossing along the shared road, it means the shared road in front of this location is congested.

Accordingly, the following interactions between mixed traffic are controlled by the traffic control signals in the next adjacent level crossing. The urban rail-bound transport vehicle U_i has to wait for the next green light phase of the traffic control signals at the next adjacent level crossing. It can start again to run along the shared road last for the duration of the green light phase. This process will be iterative until the urban rail-bound transport U_i exits the shared road.

6.4.5   Algorithm for a Whole Investigated Area

The developed algorithm can be used not only for the derivation of the waiting time function (6-11), but also for the assessment of the significances of the road traffic influences of the mixed traffic zones in a whole investigated area. Generally, a whole investigated area is not only one mixed traffic zone with one level crossing or one shared road. It can be a large-scale network of urban rail-bound transport with one or more mixed traffic zones in the whole investigated area. Therefore, the assessment of the significances of the road traffic influences is very important and necessary as a preliminary estimation as to whether or not an investigated mixed traffic zone is significantly impacted by road traffic enough to be considered as a bottleneck in the whole investigated area. Some mixed traffic zones in the investigated area are not so important and lead to a very small impact. It is meaningful to balance the accuracy of the results and the computation costs through the assessment with this algorithm.

The developed algorithm described in the previous subchapters is focused on one of the investigated mixed traffic zones in a whole investigated area. The parameter of the throughput capacity with road traffic influences $(\widehat{DS\,LF}(MTZ_i))$[19] as well as the parameter of the starting point (waiting time) of the waiting time function (6-11) $(d(MTZ_i))$ for each mixed traffic zone (MTZ_i) in the whole investigated area can be calculated with the developed algorithm.

The scheduled traffic flows of the urban rail-bound transport at each of the mixed traffic zones in the whole investigated area for an investigated time period are defined in the scheduled timetable based on the given operating program. Compared to the corresponding parameters of the throughput capacity with road traffic influences of each mixed traffic zone $(\widehat{DS\,LF}(MTZ_i))$ in the whole investigated area, the mixed traffic zone with the least increase of the traffic flow of urban rail-bound transport can be regarded as the most affected node that may be a bottleneck in the whole investigated area. Furthermore, the parameter of the throughput capacity with road traffic influences for a whole investigated area $(\widehat{DS\,LF}(WIA))$ can be determined based on the most affected node in the whole investigated area.

The parameter of the starting point of the waiting time function (6-11) for the whole investigated area $(d(WIA))$ can be derived from the parameter of the starting point of the waiting time function (6-11) at each of the mixed traffic zones $(d(MTZ_i))$ in the whole investigated area and the corresponding recovery times $(REC(MTZ_i))$ in the scheduled timetable. Consequently, the waiting time function of the whole investigated area with consideration of the road traffic influences can be approximately derived.

In conclusion, the improved waiting time function adapted to the mixed traffic zone with road traffic influences can be approximately determined with the model of this algorithm in an event-driven system. Compared to the two approaches—the modeling approach and the distribution approach described in previous Chapters 4 and 5, although the waiting time function using the event-driven system can be derived ap-

---

[19] MTZ_i: The mixed traffic zone in a whole investigated area with the index i.

proximately based on the parameters of the waiting time function without road traffic influences, it can reduce part of the efforts on data collection and the behavior of the road traffic can be reflected more exactly.

Furthermore, the developed algorithm in this chapter can be used to assess the significances of the road traffic influences in the mixed traffic zones in a whole investigated area, which can help to reduce the high costs and time by not considering the low impacted mixed traffic zones in research for a large-scale investigated area. As a preliminary estimation, the algorithm is meaningful to be carried out not only to determine the adapted waiting time function (6-11) for a single mixed traffic zone with consideration of road traffic influences, but also for the modeling approach and distribution approach beforehand.

7 Comparison of the Approaches and Validation of the Results

The results of the capacity research in urban rail-bound transport with road traffic influences are the throughput capacity and the derived recommended area of traffic flow using the adapted waiting time function described in Chapter 6. Capacity research using the simulation method has already been carried out for various research projects in pure railway systems (such as railway nodes, heavy rail, and light rail) without road traffic influences. The results are plausible and have been validated. In this research project, two different approaches, the modeling approach and the distribution approach, were discussed for capacity research. Therefore, it is meaningful to compare the results of the capacity research with the implementation of these two different approaches. Accordingly, the comparison of the impacts of the two approaches will be shown in various application areas.

7.1 Analysis of the Capacity Research Results with Various Approaches

With the modeling approach, the results of the capacity research for the investigated example are derived. The throughput capacity with road traffic influences is 104.3 trains/h for the investigated example (see Appendix Figure 1 and Appendix Table 3), which is almost the same number as with the distribution approach of 103.1 trains/h. Comparably, it can be seen that for the investigated example the throughput capacity of the railway system without the road traffic influences is obviously higher than that of where the road traffic influences in mixed traffic zones are considered. That means the potential capacity loss of urban rail-bound transport at mixed traffic zones is due to the influences of the road traffic.

Moreover, the derived recommended area of traffic flow with the adapted waiting time function (6-11), through the modeling approach with the upper limit of 91.7 trains/h for the investigated example, can be regarded as equal to the upper limit of the result through the distribution approach. The lower limit of 60.3 trains/h derived by the modeling approach differs from that of the distribution approach by about 3 trains/h less, which is a small difference. The corresponding results of the capacity research for this investigated example are shown in Table 6-6. Similarly, the upper limit of the recommended area of traffic flow with consideration of road traffic influences is nearly equal to and even lower than the lower limit of the recommended area of traffic flow without road traffic influences.

Application Case		Throughput Capacity	Recommended Area of Traffic Flow
Case without Road Traffic Influences		146.3 trains/h	93.9 – 139.8 trains/h
Case with Road Traffic Influences	Modeling Approach	104.3 trains/h	60.3 – 91.7 trains/h
	Distribution Approach	103.1 trains/h	63.6 – 92.2 trains/h

Table 6-6: Results of the Cases with and without Road Traffic Influences with the Two Approaches

By comparing Figure 4-2 and Figure 5-3, the data points from the distribution approach are more divergent than those from the modeling approach. It is understandable that the introduced perturbations are generated randomly. The random perturbations are further involved in the stochastic timetables with stepwise-varied traffic flows. The double randomness will lead to a higher variance of data points. Furthermore, due to the perturbation for the road traffic influences derived from the high traffic flow period (60 trains/h), the waiting time of the urban rail-bound transport will be relatively higher than that of the modeling approach when the traffic flow is lower since the probability of the perturbed urban rail-bound transport should be lower during the low traffic flow period with a low traffic flow of urban rail-bound transport.

7.2　Applicable Conditions

The two approaches, the modeling approach and the distribution approach, are developed for capacity research on urban rail-bound transport with road traffic influences in mixed traffic zones. In addition, the derived algorithm modeled with the event-driven system is used for determining the adapted waiting time function (6-11). Furthermore, it can be also utilized for the assessment of the significances of the road traffic influences at each mixed traffic zone in a whole investigated area.

The two approaches are both suitable for various application conditions. For a relatively simple and small-scale mixed traffic zone level crossing in an investigated area, it is suitable to use the modeling approach to collect the required data and build the simulation model through modeling the road traffic as trains in the simulation tool.

However, for the mixed traffic zone of shared roads, the behaviors of the mixed traffic are hard to be modeled for simulation. With the distribution approach, it is necessary to have statistical data of delays (waiting times) of urban rail-bound transport collected from real operation. For a complex investigated area with mixed traffic zone(s), it is unnecessary to recognize the specific conditions of each mixed traffic zone. It can be used comprehensively for various types of mixed traffic zones in a large-scale investigated area with a sufficient data base.

It is valuable to assess the significances of the road traffic influences at each of the mixed traffic zones. Consequently, the superfluous efforts can be avoided with the developed algorithm based on the event-driven system to evaluate the mixed traffic zones in an investigated area. This algorithm can be used for the basic assessment of the subsequent execution of capacity research with consideration of mixed traffic. Additionally, if the required information is not fully available, the developed algorithm for the mixed traffic model can be used to derive the adapted waiting time function (6-11) approximately with the value of the starting point d $(= d_5/(1 - \eta)^{2/3})$ and the throughput capacity with road traffic influences $\widehat{DS\,LF}$, as the results of capacity research for further evaluation.

8 Summary

In urban mixed traffic zones, the urban rail-bound transport is heavily influenced by road traffic. These influences cannot be ignored, as they lead to a great amount of loss to the urban rail-bound transport capacity. It is meaningful to evaluate the potential loss of capacity due to road traffic influences. The interactions between the urban rail-bound transport and road traffic may result in possible economic costs, not only for the urban rail-bound transport but also for the road traffic. A standardized procedure to evaluate possible improvement measures (e.g. to build new tunnels or bridges to separate urban rail-bound transport and road traffic) can be further developed.

The two approaches developed in this research can be used to carry out capacity research with consideration of the road traffic influences on urban rail-bound transport. With these two approaches, the stochastic influences caused by road traffic are described systematically. According to the required application conditions, it is necessary to select a suitable approach. For the modeling approach, the regulations of the traffic control signaling system reflect the operation of "modeled road traffic" at level crossings during the investigated time period. It is suitable to use a modeling approach for relatively simple mixed traffic zones in an investigated area. On the other hand, the distribution approach is applicable for a complex mixed traffic zone in an investigated area with sufficient statistical data of delays to determine the perturbations for the road traffic influences, which are introduced into the simulation model to represent the road traffic influences in the results of the capacity research.

For capacity research, a key ingredient is the waiting time function, which can be derived from simulations of the stochastically-influenced timetables with stepwise-varied traffic flows. The adapted waiting time function plays a very important role to fit the discrete waiting time points and then to derive the recommended area of traffic flow for the situation of urban mixed traffic. The waiting time function with a higher (adjusted) coefficient of determination can represent the trend of the discrete waiting times better. The improved waiting time function has a better (adjusted) coefficient of determination in comparison to the existing one. The fitted curve of the discrete waiting time with the newly adapted waiting time function shows results much closer to reality. Accordingly, the derived recommended area of traffic flow is much more plausible.

The preliminary derivation of the throughput capacity with road traffic influences $\widehat{DS\,LF}$ and the parameter d of the adapted waiting time function allow for the use of the developed algorithm to determine the results of the capacity research. Moreover, it is also possible to reduce the costs of the following work for capacity research with this developed algorithm by assessing the various significances of the road traffic influences at each of mixed traffic zone in an investigated area.

This research project has been carried out within the given schedule. As part of the research project, a two-day, project-specific workshop that was funded by DFG took place. It was a platform for the information exchange on relevant topics of the state of the research, specifically regarding the methodologies to evaluate and increase the capacity of railway networks and discussing practical applications and further development. Within the framework of this research project, two student theses ([Morman 2014] und [Paracha 2015]) related to this research project were accomplished. During the research work, a scientific paper for publication with partial results was completed and published with the topic "Einfluss des Straßenverkehrs auf die Wartezeitfunktion bei Leistungsuntersuchungen im SPNV" (see [Martin & Di Liu 2016]).

Abbreviations

Compatible Routes Combination	CRC
East	E/ E'
Exclusive Routes Combination	ERC
High Traffic Flow Period	HTP
Location Point	LP
Low Traffic Flow Period	LTP
North	N/ N'
Normal Traffic Flow Period	NTP
Programm zur Untersuchung des Leistungsverhaltens von Eisenbahninfrastrukturen	PULEIV
Road Traffic	ROT
South	S/ S'
Urban Rail-bound Transport	URT
Verkehr In Städten – SIMulationsmodell	VISSIM
West	W/ W'

Formula Symbols

$a,\ b,\ a_i,\ b_i,\ c_i,\ d_i$	Parameters of the waiting time function
α	The nonlinear parameters of the model function $\hat{y}_i(x)$
β	The linear parameters of the model function $\hat{y}_i(x)$
$B_{bef}(\eta)$	Traffic energy function of the waiting time
$deg.$	The degrees of freedom of the waiting time function
del_m	Average value of delays in minutes
del_{max}	Maximum delay for perturbed urban rail-bound transport in minutes
ϕ_j	The j th of n basis function of x
EL_v	Average number of requests in queuing system
ET_a	Expectancy value of time interval between the arrival times of two trains
ET_b	Expectancy value of the service time in the server of the queuing system
ET_B	Average transport time
$ETw(\eta)$	(statistical) Expectancy value of the waiting time function
η	Utility Factor of Capacity
λ	Average arrival frequency (rate)
μ	Average service rate of a single server
ρ	Utility factor
$GP_{ROT,p}$	The green light phase of the road traffic route $r_{ROT,p}$
h	The number of times that the urban rail-bound transport is hindered along the shared road
hl_i	The leverages that adjust the residuals by reducing the

	weight of high leverage data points
$h_{URT,p}$	The hindrance time of urban rail-bound transport route caused by road traffic route $r_{ROT,p}$ in one traffic cycle at an investigated mixed traffic zone
$h_{URT,q,p}$	The effective hindrance time of an urban rail-bound transport route $r_{URT,q}$ caused by road traffic route $r_{ROT,p}$
H_{URT}	The hindrance time of urban rail-bound transport caused by road traffic
K	A tuning constant equal to 4.685
l	The distance from the location point (LP_l), at which the urban rail-bound transport is first hindered by road traffic to the end point of the shared road
LP_l	Defined location point along the shared road with the index l
L_{ROT}	The lines of the rail tracks for trains of the "modeled road traffic"
L_{URT}	The lines of the rail tracks for the "trains" of urban rail-bound transport
m	The number of road traffic routes with $|R_{ROT}| = m$
MAD	The median absolute deviation of the residuals
$MS(Res)$	The residual mean of squares
$MS(Total)$	The total mean of squares
MTZ_i	The mixed traffic zone in a whole investigated area with the index i
n	The number of the urban rail-bound transport routes with $|R_{URT}| = n$
n_{req}	Possible number of requests in the queuing system

n_s	The number of sample data points
N	Traffic flow (trains/h)
N_{req}	The number of requests in the queuing system
$o_{ROT,p}$	The effective occupation time of the road traffic route $r_{ROT,p}$
O_{ROT}	The potential occupation time for the road traffic
$pro.$	Proportion values of perturbed urban rail-bound transport (delayed urban rail-bound transport) in %
P_{ROT}	The perturbation parameters for the running time extension caused by road traffic influences
P_{URT}	The perturbation parameters for the running time extension without road traffic influences
r_{adj}	The usual least squares residuals
res_i	The residual for the i th data point
R^2	Coefficient of determination
$\bar{R}^2$	Adjusted coefficient of determination
$r_{ROT,p}$	The road traffic route with the index $p \in [1, m]$
R_{ROT}	The set of routes of road traffic
$r_{URT,q}$	The urban rail-bound transport route, with the index $q \in [1, n]$
R_{URT}	The set of routes of urban rail-bound transport
$RP_{ROT,p}$	The red light phase of the road traffic route $r_{ROT,p}$
S	The length of the shared road
s_i	Traffic control signal with the number i
s_{rob}	The robust variance given by $MAD/0{,}6745$

$S_{rel}(\eta)$	Relative sensitivity of the waiting time function
SS	The summed square of residuals where m is the number of data points included in the fit
$SS(Res)$	The residual sum of squares
$SS(Total)$	The total sum of squares
$S_{W,ROT/URT}, S_{E,ROT/URT}, S_{N,ROT/URT}, S_{S,ROT/URT}$	The stops for the mixed traffic with directions of the investigated example
t	The index of the investigated time period, $t \in [1,\ T]$
T	The investigated time period in hours
$DS\ LF$	Throughput capacity
$\widehat{DS\ LF}$	Throughput capacity with road traffic influences
T_{sl}	The time slice in seconds
u_i	The standardized adjusted residuals of i th iterative
U_i	A train of urban rail-bound transport on its corresponding route $r_{URT,q+1}$ with the index i
v	The velocity of urban mixed traffic (road traffic)
v_{bef}	Average speed of trains
w_i:	The robust bisquare weights $w_{bisqure}$ of the i th iterative
x_i	The variables of the model function $\hat{y}_i(x)$
y_i	The i th of m fitted values with the model function
$\hat{y}_i$	The i th of m fitted values with the model function
$YP(G)_{ROT,p}$	The yellow light phase of the road traffic $r_{ROT,p}$ changed from its green light phase
$YP(R)_{ROT,p}$	The yellow light phase of road traffic $r_{ROT,p}$ changed from its red light phase

Glossary

Cycle time	Cycle time is the time duration that it takes to complete one traffic signal cycle at a level crossing.
Effective occupation time of road traffic route	Effective occupation time of road traffic route is the time duration that it takes to complete one road traffic route ($r_{ROT,p}$) starting from its entry until another exclusive mixed traffic route (road traffic or urban rail-bound transport route) is allowed to enter the mixed traffic zone.
Effective hindrance time of urban rail-bound transport route	The effective hindrance time of an urban rail-bound transport route is the time duration that an urban rail-bound transport vehicle is hindered by the occupation of road traffic at an investigated mixed traffic zone.
Hindrance time of urban rail-bound transport caused by road traffic	The hindrance time of urban rail-bound transport caused by road traffic is the total occupation time of the road traffic, at which the urban rail-bound transport vehicles are hindered by the occupation of the road traffic. It is a time interval that is the maximum waiting time of one or more hindered urban rail-bound transport vehicles in the mixed traffic zone that is occupied by one road traffic route.
Instantaneous values of indicators	Instantaneous values are the values of indicators (del_m and $pro.$) that are the simulation results of output delays related to the given perturbation parameters.
Investigated area	Investigated area is a defined part of the railway infrastructure on which the capacity research will be carried out.
Level crossing	A level crossing is an intersection where an urban

rail-bound line crosses an urban roadway or path at the same level. The term also applies when an urban rail-bound line with separate right-of-way or reserved track crosses a road in the same manner.

Maximum capacity

Maximum capacity is a theoretical value and doesn't meet the requirements of operational quality. It allows for an unlimited congestion situation without keeping the structure of the operating program. It corresponds to the theoretical maximum number of trains and shunting movements on the investigated infrastructure within a given time period [DB Netz AG 2008].

Modeled road traffic

Modeled road traffic are the trains representing the road traffic at the investigated mixed traffic zone modeled in the simulation tool RailSys for the modeling approach.

Operating program

Operating program is the comprehensive description of the performance and requirements of the railway operation, including the number of train runs, the train properties, their structure and sequence, as well as the temporal allocation of the train runs [DB Netz AG 2008].

Perturbation

Perturbations are the disruptions to the scheduled timetable during operations. They are introduced into the simulation model for rail-bound transport in the scheduled timetable.

Perturbation parameters

Perturbation parameters can be used to describe the distribution of one perturbation in the simulation model. There are three perturbation parameters: average delay, proportion values of perturbed urban rail-bound transport, and maximum delay for delayed

	urban rail-bound transport.
Perturbation type	There are three types of perturbations studied in this research project: initial delay, dwell time extension, and running time extension.
Potential occupation time for road traffic O_{ROT}	The potential occupation time for road traffic is the maximal possible occupation time of the road traffic at an investigated mixed traffic zone in an investigated time period. It is a time interval that is maximally allocated to the exclusive use of road traffic route at an investigated mixed traffic zone in an investigated time period.
Recommended area of traffic flow	Recommended area of traffic flow is the range of traffic flow to reach a customer-friendly quality of operation with the optimum utilization of a given infrastructure. ([Hertel 1992], [Schmidt 2009] & [Chu 2014]).
Recovery time	Recovery time is a time supplement that is added to the pure running time and dwell time to enable a train to make up small delays [Pachl 2014].
Road traffic	Road traffic refers to all traffic types moving on the urban roadway, which can be further categorized into motorized road traffic and non-motorized road traffic. Motorized road traffic includes private vehicles such as trucks, private cars, and motor (-bicycles); and public vehicles such as busses. Non-motorized road traffic includes pedestrians and bicycles.
Route	A route is defined as a section of infrastructure for urban rail-bound transport or an urban roadway for road traffic with a specific direction, on which the mixed traffic can move from the entry point A to exit point B.

Shared road	A shared road is one basic type of mixed traffic zone where the rail tracks for urban rail-bound transport share the urban roadways and even the bikeways (bicycle lanes) with the road traffic.
throughput capacity	Throughput capacity is the average traffic flow of all possible maximum traffic flows per time unit (hour) in the static phase of the operating procedure with various train sequences of a defined rough operating program (train mixture). The throughput capacity (incoming traffic flow = outgoing traffic flow) has to keep the structure of the defined operating program (train mixture) on a given infrastructure.[Chu 2014].
Traffic flow	Traffic flow is the number of trains or train paths during a given investigation period within the investigated area [DB Netz AG 2008]
Target values of indicators	Target values of indicators are the values of indicators (del_m and $pro.$) that are given from collected operational data in reality and are used as reference values for calibrating a model.
Traffic light phase	A traffic light phase is defined as the time duration to direct the movements of road traffic controlled by the corresponding traffic control signals. It gives the indication for movements of road traffic with certain time duration. There are four traffic light phases discussed in this research project, the red light phase, the green light phase, and two yellow light phases changed from either the red light or the green light.
Traffic control signal	A traffic control signal is also known as a traffic light, which is a signaling device positioned at intersections, pedestrian crossings, and other locations to

	control the behaviors of traffic loads, whether to move or be still with corresponding indications of the traffic light.
Train mixture	Train mixture is the structure of the operating program, which is the rough operating program including the characteristics of the various train groups in the model and the ratio of the number of trains in each train group.
Urban mixed traffic	The different types of transport that operate in urban mixed traffic zones including urban rail-bound transport are collectively called urban mixed traffic. It can be categorized into public transport (urban rail-bound transport and bus) and private transport (individual vehicle and pedestrian) or into urban rail-bound transport and road traffic that is mainly used in this research project.
Urban mixed traffic zone	An urban mixed traffic zone is an urban area in which not only urban rail-bound transport but also mutual interfered other transports (urban mixed traffic) operate.
Urban rail-bound transport	Urban rail-bound transport is the rail-bound system that operates in urban areas and includes the metro, light rail transit, and trams.

Bibliography

[Anderson et al. 2011] *Anderson, D. R.; Sweeney, D. J.; Williams, T. A.*: Statistics for Business and Economics. South - Western Cengage Learning (2011).

[Blower & Dowlatabadi 1994] *Blower, S. M.; Dowlatabadi, H.*: Sensitivity and Uncertainty Analysis of Complex Models of Disease Transmission: An HIV Model, as an Example. International Statistical Review / Revue Internationale De Statistique 62(2) (1994), S. 229–243.

[Blue et al. 1997] *Blue, V. J.; Embrechts, M. J.; Adler, J. L.*: Cellular Automata Modeling of Pedestrian Movements. IEEE (1997).

[Bosse et al. 1995] *Bosse, G.; Martin, U.; Pachl, J.*: Anwendung des Simulationsprogramms UX - SIMU zur Leistungsuntersuchung von Strecken. In: Schriftenreihe des Instituts für Verkehrssystemtheorie und Bahnverkehr Nr. 1 (1995), S. 105–126.

[Chu 2014] *Chu, Z.*: Modellierung der Wartezeitfunktion bei Leistungsuntersuchungen im Schienenverkehr unter Berücksichtigung der transienten Phase. Dissertation, Universität Stuttgart, Institut für Eisenbahn- und Verkehrswesen. Betreut von Ullrich Martin. Schriftenreihe VWI Neues verkehrswissenschaftliches Journal - Ausgabe 10. Norderstadt: BoD - Book on Demand., 2014.

[Cui et al. 2014] *Cui, Y.; Ullrich, M.; Zhao, W.; Cao, N.*: Entwicklung eines Algorithmus für die Kalibrierung von Modellen zur Betriebssimulation in spurgeführten Verkehrssystemen unter Berücksichtigung stochastischer Bedingungen. Deutsche Forschungsgemeinschaft (DFG) - Forschungsvorhaben (GZ: MA 2326/9-1). Schriftenreihe VWI Neues verkehrswissenschaftliches Journal - Ausgabe 9. Norderstadt: BoD - Book on Demand. (2014).

[Czitrom 1999] *Czitrom, V.*: One-Factor-at-a-Time Versus Designed Experiments. The American Statistician 53, No. 2 (1999).

[DB Netz AG 2008] *DB Netz AG*: DB Richtlinie 405 - Fahrwegkapazität (2008).

[Draper & Smith 1998] *Draper, N. R.; Smith, H.*: Applied regression analysis. Wiley, New York, Chichester, 1998.

[Everitt 2002] *Everitt, B.*: The Cambridge dictionary of statistics. Cambridge University Press, Cambridge, 2002.

[Fellendorf & Vortisch 2010] *Fellendorf, M.; Vortisch, P.*: Microscopic Traffic Flow Simulator VISSIM. In: Barceló, J. (Hrsg.): Fundamentals of Traffic Simulation. Springer New York, Editor: Barceló, Jaume, New York, NY, 2010, S. 63–93.

[Fukui & Ishibashi 1996] *Fukui, M.; Ishibashi, Y.*: Traffic Flow in 1D Cellular Automaton Model Including Cars Moving with High Speed. Journal of the Physical Society of Japan 65 (1996), 6, S. 1868–1870.

[Gipps 1981] *Gipps, P. G.*: A behavioural car-following model for computer simulation. Transportation Research Part B: Methodological 15 (1981), 2, S. 105–111.

[Green 1984] *Green, L.*: A Queueing System with Auxiliary Servers. Management Science 30 (1984), S. 1207–1216.

[Hamby 1995] *Hamby, D. M.*: A comparison of sensitivity analysis techniques. Health physics 68 (1995), pp. 195–204.

[Hertel et al. 1987] *Hertel, G.; Ludwig, D.; Bauer, J.*: Leistung und Qualität im Eisenbahntransport. In: Wissenschaftliche Zeitschrift Hochschule für Verkehrswesen Dresden Nr. 2, Jg. 34 (1987), S. 207–234.

[Hertel 1992] *Hertel, G.*: Die maximale Verkehrsleistung und die minimale Fahrplanempfindlichkeit auf Eisenbahnstrecken. In: ETR-Eisenbahntechnische Rundschau Nr. 10, Jg. 41. (1992), S. 665–671.

[International Union of Railways (UIC) 2004] *International Union of Railways (UIC)*: UIC CODE 406: Capaciy, 2004.

[Janecek et al. 2010b] *Janecek, D.; Weymann, F. H. G.; Schaer, T.*: LUKS - integriertes Werkzeug zur Leistungsuntersuchung von Eisenbahnknoten und -strecken. In: ETR-Eisenbahntechnische Rundschau 59 (1+2) (2010b), S. 25–32.

[Janecek & Weymann 2010a] *Janecek, D.; Weymann, F. H. G.*: LUKS - Analysis of lines and junctions. In: 12th World Conference on Transport Research, Instituto Superior Tecnico, Lisboa, 2010a.

[Kontaxi & Ricci 2010] *Kontaxi, E.; Ricci, S.* (Hrsg.): Capacity Analysis: Methodological Framework and Harmonization Perspectives, University of Rome, DICEA Department, Italy, 2010.

[Li 2015] *Li, X.*: Mikroskopische Engpassanalyse bei eisenbahnbetriebswissenschaftlichen Leistungsuntersuchungen. Dissertation, Universität Stuttgart, Institut für Eisenbahn- und Verkehrswesen. Betreut von Ullrich Martin. Schriftenreihe VWI Neues verkehrswissenschaftliches Journal - Ausgabe 14. Norderstadt: BoD - Book on Demand., 2015.

[Lindner 2011] *Lindner, T.*: Applicability of the Analytical Compression Method for Evaluating Node Capacity. Universität Braunschweig, Institut für Eisenbahnwesen und Verkehrssicherung IfEV (2011).

[Liu 2011] *Liu, D.*: Microscoic Capacity Research and Performance Evaluation of Operation Quality in Ulm Main Station. Master thesis, Universität Stuttgart, Institut für Eisenbahn- und Verkehrswesen, 2011.

[Ludwig 1990] *Ludwig, D.*: Beitrag zur Leistungs- und Qualitätssicherung von Streckenfahrplä-nen der Eisenbahn. Dissertation, Dresden, 1990.

[Martin et al. 2011] *Martin, U.; Schmidt, C.; Chu, Z.*: PULEIV Anwenderleitfaden. zur PULEIV - Version 2.1. Anleitung zur Ermittlung des Leistungsverhaltens von Eisenbahninfrastrukturen mit Hilfe von Simulationsprogrammen., Stuttgart, 2011.

[Martin et al. 2012] *Martin, U.; Chu, Z.; Peter Breuer*: Leistungsuntersuchung Staatsgalerie im Rahmen der Umbaumaßnahmen zu Stuttgart 21. Abschlussbericht (unpublished), Im Auftrag der Stuttgarter Straßenbahnen AG, 2012.

[Martin 2014] *Martin, U.* (Hrsg.): Railway Timetabling & Operations: 12. Performance Evaluation. Chapter 9: Simulation. Eurailpress in DVV Media Group. Editors: Hansen, Ingo A.; Pachl, Jörn, 2014.

[Martin & Breuer 2008] *Martin, U.; Breuer, P.*: Mögliche Trassenlagen der Regionalzüge auf der Gäubahn, Ergänzende Betrachtung zur Leistungsuntersuchung der Station Terminal. Abschlussbericht (unpublished), Im Auftrag von DB Projekt GmbH, 2008.

[Martin & Chu 2012] *Martin, U.; Chu, Z.*: Dynamisierung von Zeitscheiben in Betriebsprogrammen bei Leistungsuntersuchungen. In: ETR-Eisenbahntechnische Rundschau Nr. 5 (2012), S. 40–45.

[Martin & Chu 2013] *Martin, U.; Chu, Z.*: Direkte Experimentelle Bestimmung der Maximalen Leistungsfähigkeit bei Leistungsuntersuchungen im spurgeführten Verkehr. Deutsche Forschungsgemeinschaft (DFG) - Forschungsvorhaben (GZ: MA 2326/6-1). Schriftenreihe VWI Neues verkehrswissenschaftliches Journal-NVJ Ausgabe 7, Norderstadt: BoD - Books on Demand (2013).

[Martin & Di Liu 2016] *Martin, U.; Di Liu*: Einfluss des Straßenverkehrs auf die Wartezeitfunktion bei Leistungsuntersuchungen im SPNV. In: ETR-Eisenbahntechnische Rundschau Nr. 1+2 (2016), S. 21–25.

[Mckay et al. 2000] *Mckay, M. D.; Conover, W. J.; Beckman, R. J.*: A comparison of three methods for selecting values of input variables in the analysis of output from a computer code. Technometrics Vol. 42, No. 1 Special 40th Anniversary Issue (2000), S. 55–61.

[Michl & Sojka 2014] *Michl, Z.; Sojka, M.*: Microsimulation as a tool for evaluation of infrast ructure and operational concept alternatives in a complex railway node. in: Logistyka 4 (2014), S. 3029–3037.

[Moler 2008] *Moler, C. B.*: Numerical computing with MATLAB. Society for Industrial and Applied Mathematics (SIAM), Philadelphia, ©2004, 2008.

[Moré & Sorensen 1983] *Moré, J. J.; Sorensen, D. C.*: Computing a Trust Region Step. SIAM Journal on Scientific and Statistical Computing 4 (1983), S. 553–572.

[Morman 2014] *Morman, B.*: Untersuchung der Störungsverteilungen entsprechend dem Mischverkehr am Berliner Platz in Stuttgart. Bachelorarbeit, Universität Stuttgart, Institut für Eisenbahn- und Verkehrswesen. Betreut von Di Liu, 2014.

[Nagel & Schreckenberg 1992] *Nagel, K.; Schreckenberg, M.*: A Cellular Automaton Model for Freeway Traffic. Journal de Physique I Vol. 2 (12) (1992), S. 2221–2229.

[Newell 2002] *Newell, G. F.*: A simplified car-following theory: A lower order model. Transportation Research Part B: Methodological 36 (2002), S. 195–205.

[Nießen 2008] *Nießen, N.*: Leistungskenngrößen für Gesamtfahrstraßenknoten. Dissertation, Aachen, 2008.

[Okais et al. 2010] *Okais, C.; Roche, S.; Kurzinger, M.-L.; Riche, B.; Bricout, H.; Derrough, T.; Simondon, F.; Ecochard, R.*: Methodology of the sensitivity analysis used for modeling an infectious disease. Vaccine Vol. 28 (51) (2010), pp. 8132–8140.

[Omahen 1977] *Omahen, K. J.*: Capacity Bounds for Multiresource Queues. J. ACM (Journal of the Association for Computing Machinery) 24 (1977), S. 646–663.

[Pachl 2014] *Pachl, J.*: Railway Operation and Control. VTD Rail Publishing, Mountlake Terrace (USA), 2014.

[Pachl 2016] *Pachl, J.*: Systemtechnik Des Schienenverkehrs: Bahnbetrieb Planen, Steuern Und Sichern. Vieweg + Teubner Verlag, 2016.

[Paracha 2015] *Paracha, M. B.*: Investigation on Clustering Mixed Traffic Zones with Individual Transport Influences related to Capacity of Urban Rail-bound Transport. Master thesis, Universität Stuttgart, Institut für Eisenbahn- und Verkehrswesen. Betreut von Di Liu, 2015.

[Potthoff 1969] *Potthoff, G.*: Die Bedienungstheorie im Verkehrswesen. Transpress, VEB Verlag für Verkehrswesen Berlin (1969).

[Potthoff 1972] *Potthoff, G.*: Verkehrsströmungslehre. VEB Verlag für Verkehrswesen Berlin (1972).

[Rickert et al. 1996] *Rickert, M.; Nagel, K.; Schreckenberg, M.; Latour, A.*: Two lane traffic simulations using cellular automata. Physica A: Statistical Mechanics and its Applications 231 (1996), S. 534–550.

[RMCon 2010] *RMCon*: Manual/ Handbuch RailSys 7, 2010.

[Rossetti 2009] *Rossetti, M. D.*: A Simulation-based Approach for Estimating the Commercial Capacity Railways, IEEE, 2009.

[Saltelli et al. 2008] *Saltelli, A.; Ratto, M.; Andres, T.; Campolongo, F.; Cariboni, J.; Gatelli, D.; Saisana, M.; Tarantola, S.*: Global Sensitivity Analysis. The Primer. Biometrics 65 (2008), S. 1311–1312.

[Saltelli et al. 2010] *Saltelli, A.; Annoni, P.; D'Hombres, B.*: How to avoid a perfunctory sensitivity analysis. Procedia - Social and Behavioral Sciences 2 (2010), S. 7592–7594.

[Schmidt 2009] *Schmidt, C.*: Beitrag zur experimentellen Bestimmung der Wartezeitfunktion bei Leistungsuntersuchungen im spurgeführten Verkehr. Dissertation, Universität Stuttgart, Institut für Eisenbahn- und Verkehrswesen, 2009.

[Schwanhäußer 1978] *Schwanhäußer, W.*: Die Ermittlung der Leistungsfähigkeit von großen Fahrstraßenknoten und von Teilen des Eisenbahnnetzes. Archiv für Eisenbahntechnik 33 (1978), S. 7–18.

[Siefer 2014] *Siefer, T.* (Hrsg.): Railway Timetabling & Operations: 9. Simulation. Eurailpress in DVV Media Group. Editors: Hansen, Ingo A.; Pachl, Jörn, 2014.

[Sorensen 1982] *Sorensen, D. C.*: Newton's Method with a Model Trust Region Modification. SIAM Journal on Numerical Analysis 19 (1982), 2, S. 409–426.

[Venables & Ripley 1997] *Venables, W. N.; Ripley, B. D.*: Robust Statistics. In: Chambers, J.; Eddy, W.; Härdle, W.; Sheather, S.; Tierney, L.; Venables, W. N.; Ripley, B. D. (Hrsg.): Modern Applied Statistics with S-PLUS. Springer New York, New York, NY, 1997, S. 247–266.

[Wakob 1985] *Wakob, H.*: Ableitung eines generellen Wartemodells zur Ermittlung der planmäftigen Wartezeiten im Eisenbahnbetrieb unter besonderer Berücksichtigung der Aspekte Leistungsfähigkeit und Anlagenbelastung. verkehrswissenschaftlichen Institutes der RWTH Aachen 36 (1985).

[Wendler 2002] *Wendler, E.*: Bemessungsmethoden für große Eisenbahnknoten Effizienter Bahnbetrieb. In: ETR-Eisenbahntechnische Rundschau 51(7/8) (2002), S. 418–424.

[Wolfram 1986] *Wolfram, S.*: Theory and Applications of Cellular Automata, World Scientific, 1986.

[Yuan 2006] *Yuan, J.*: Stochastic Modelling of Train Delays and Delay Propagation in Stations. Ph.D. thesis, Delft University of Technology, 2006.

Appendix

The data shown in the following two tables (Appendix Table 1and Appendix Table 2) is collected from on-site measurements at an investigated level crossing under various time-related influential parameters—time periods of operation and days of operation in two entire weeks and two additional Saturdays, Sundays, and holidays.

Time Periods of Operation						
Time Periods	Low Traffic Flow Period (LTP)		Normal Traffic Flow Period (NTP)		High Traffic Flow Period (HTP)	
Classes of Waiting Times [s]	Number of Trains	Probability	Number of Trains	Probability	Number of Trains	Probability
0	0	64	0.215488215	83	0.161165049	60	0.102564103
>0	6	40	0.134680135	50	0.097087379	13	0.022222222
7	12	99	0.333333333	119	0.231067961	75	0.128205128
13	18	35	0.117845118	80	0.155339806	110	0.188034188
19	24	20	0.067340067	56	0.108737864	76	0.12991453
25	30	18	0.060606061	43	0.083495146	69	0.117948718
31	36	9	0.03030303	30	0.058252427	57	0.097435897
37	42	8	0.026936027	23	0.044660194	51	0.087179487
43	48	3	0.01010101	10	0.019417476	20	0.034188034
49	54	0	0	7	0.013592233	16	0.027350427
55	60	1	0.003367003	0	0	12	0.020512821
61	66	0	0	4	0.00776699	10	0.017094017
67	72	0	0	3	0.005825243	5	0.008547009
73	78	0	0	7	0.013592233	3	0.005128205
79	84	0	0	0	0	3	0.005128205
85	90	0	0	0	0	3	0.005128205
91	96	0	0	0	0	2	0.003418803
Sum		297	1	515	1	585	1

Appendix Table 1: Data Base of the Number of Trains with Corresponding Waiting time Measured Practice On-Site at the Investigated Mixed Traffic Zone of Level Crossing under Various Time Periods of Operation

Day Types of Operation							
Day Types		Mon. - Fr.		Saturday		Sunday and Holidays	
Classes of Waiting Times [s]		Number of Trains	Probability	Number of Trains	Probability	Number of Trains	Probability
0	0	157	0.153470186	103	0.278378378	111	0.299191375
>0	6	143	0.139784946	67	0.181081081	71	0.191374663
7	12	214	0.209188661	69	0.186486486	70	0.188679245
13	18	147	0.143695015	42	0.113513514	49	0.132075472
19	24	110	0.107526882	27	0.072972973	33	0.088948787
25	30	89	0.086999022	19	0.051351351	18	0.04851752
31	36	50	0.048875855	14	0.037837838	7	0.018867925
37	42	40	0.039100684	10	0.027027027	5	0.013477089
43	48	23	0.022482893	7	0.018918919	4	0.010781671
49	54	10	0.009775171	4	0.010810811	2	0.005390836
55	60	12	0.011730205	3	0.008108108	1	0.002695418
61	66	11	0.010752688	2	0.005405405	0	0
67	72	10	0.009775171	2	0.005405405	0	0
73	78	7	0.00684262	1	0.002702703	0	0
79	84	0	0	0	0	0	0
85	90	0	0	0	0	0	0
91	96	0	0	0	0	0	0
Sum		1023	1	370	1	371	1

Appendix Table 2: Data Base of the Number of Trains with Corresponding Waiting time Measured Practice On-Site at the Investigated Mixed Traffic Zone of Level Crossing under Various Days of Operation

The investigated example is an urban rail-bound transport network with the infrastructure shown in Appendix Figure 1 and the operating program shown in Appendix Table 3. The investigated mixed traffic zone level crossing described in Subchapter 4.2.1 is one part of the whole investigated area and is demarcated with a red circle. There are 25 scheduled stops with 8 modeled stop positions for the road traffic. Four of the five operation lines of urban rail-bound transport ($L_{URT,1}$ - $L_{URT,4}$) are running through the mixed traffic zone in the whole investigated area. There are in total 180 trains per hour running in the whole investigated area with 120 trains of "modeled road traffic" and 60 "trains" of urban rail-bound transport wherein 168 trains are running through the investigated mixed traffic zone level crossing. The overview of the train runs is shown in Appendix Table 3.

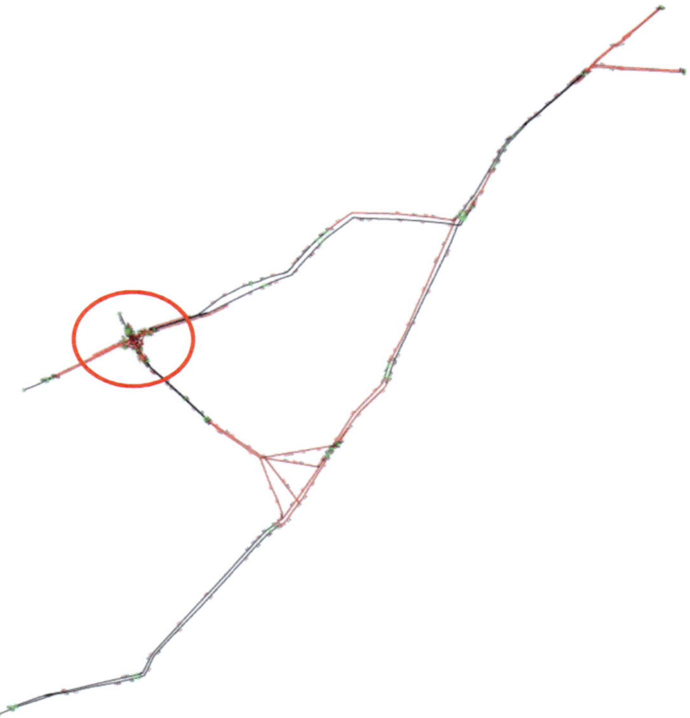

Appendix Figure 1: The Infrastructure of the Investigated Example with a Mixed Traffic Zone of Level Crossing

Operating Program		
Train Mixture		**Traffic Flow for each direction [trains/h]**
"Trains" of urban Rail-bound Transport	$L_{URT,1}$	6
		6
	$L_{URT,2}$	6
		6
	$L_{URT,3}$	6
		6
	$L_{URT,4}$	6
		6
	$L_{URT,5}$	6
		6
Trains of "modeled road traffic"	$L_{ROT,1}$	30
	$L_{ROT,2}$	30
	$L_{ROT,3}$	30
	$L_{ROT,4}$	30

Appendix Table 3: The Corresponding Operating Program of the Investigated Example